Ripley's Believe It or Not!®

Vice President, Licensing & Publishing
Amanda Joiner

Editorial Manager
Carrie Bolin

Editor Jordie R. Orlando

Junior Editor Briana Posner

Designer Shawn Biner

Proofreader Rachel Paul

Factchecker Yvette Chin

Reprographics Bob Prohaska

Cover Design Chris Conway, Luis Fuentes

Special thanks to the writers and editors of IFLScience.com and Ripleys.com

Published by Ripley Publishing 2020
under license with IFL Science

10 9 8 7 6 5 4 3 2 1

ISBN: 978-1-60991-383-0

For more information regarding permission, contact:
VP Licensing & Publishing

Ripley Entertainment Inc.
7576 Kingspointe Parkway, Suite 188
Orlando, Florida 32819
Email: publishing@ripleys.com
www.ripleys.com/books

Manufactured in China in January 2020.

First Printing

Library of Congress Control Number: 2019956371

a Jim Pattison Company

Science frequently teeters on the edge of the unbelievable, often leaving people—even scientists—thinking, "What the. . . ?" For instance, did you know that pink isn't a real color? Or that hip-hop music makes cheese taste better?

The editors of Ripley's Believe It or Not! and IFL Science have joined forces to bring you the strangest, most unbelievable science stories they can find. From historical oddities to cutting-edge technology, and strange animals to cosmic conundrums, there is no shortage of weird, true facts that will leave you scratching your head and thinking. . .

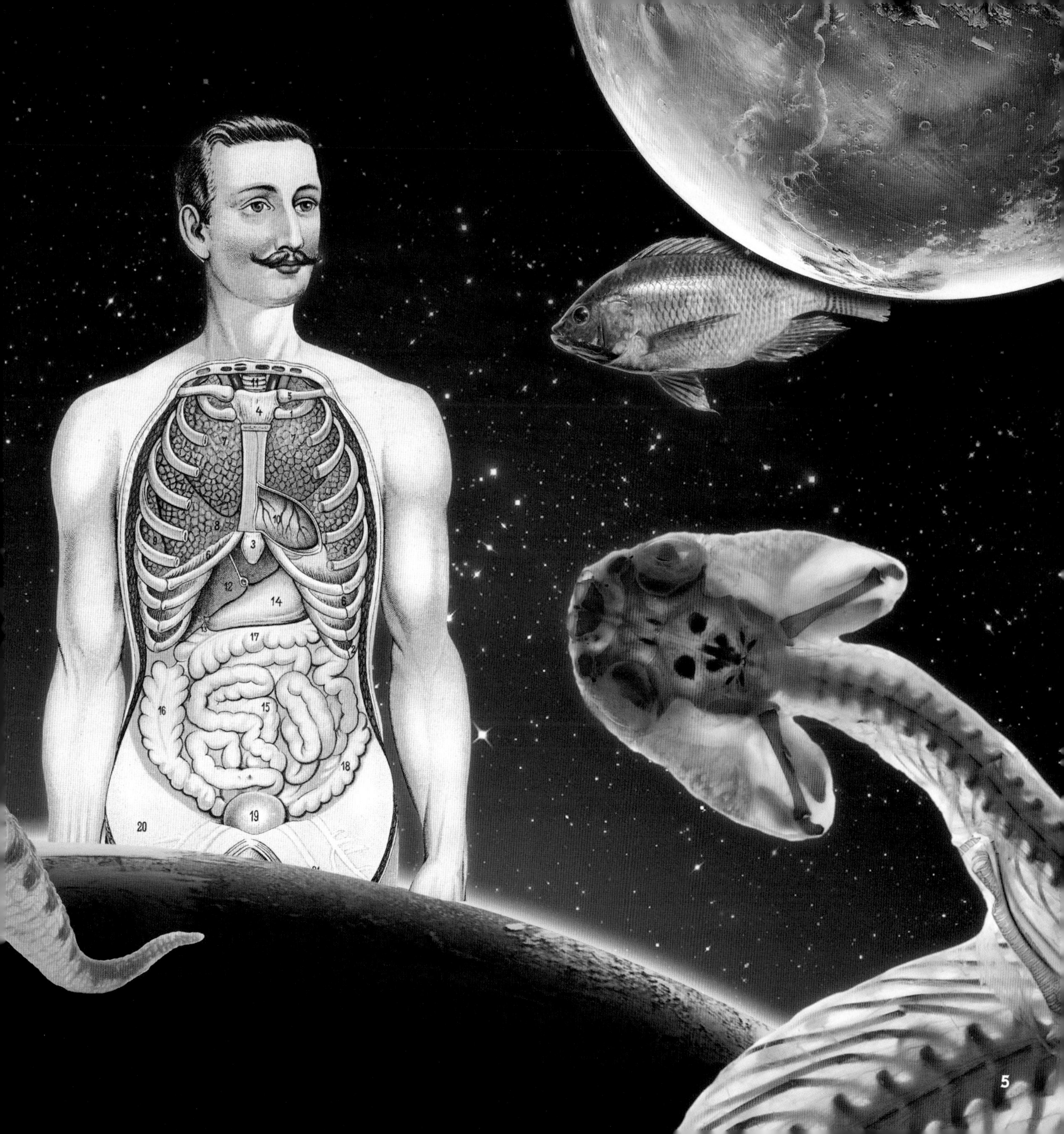

PINK ISN'T REAL

Take a close look at a rainbow and search for the color pink. Did you find it? No? That's because there is no pink wavelength of light.

When you see pink, your brain is actually piecing together red and white wavelengths. Our perception of color is not as simple as a linear spectrum; it also involves tints (added light values) and shades (added dark values). By adding tints to red, we lighten the color, as well as shift it toward the blue end of the spectrum and endow it with a different quality. We call this quality "pink."

If you want to get metaphysical about it, all "colors" are just abstractions based on interpretations of the body and mind that can't exist outside of the visual system. Don't think about it too hard; it can get pretty existential.

INDESTRUCTIBLE SPORES

Molds are present in every human environment, but the pristine vacuum of space must be perfectly clean, right? Wrong!

Mold has been found on the International Space Station (ISS). Since the ISS is technically a human environment, mold growth inside the spacecraft is a common annoyance. Astronauts spend many hours each week making sure the mold stays controlled enough so as not to become a health problem.

Mold can survive many extreme environments, but researchers have found that spores of *Aspergillus* and *Penicillium* can withstand exposure to X-rays 200 times higher than the dose that would kill a human.

Exceedingly resilient, these spores are now thought to be capable of living on the *outside* of the ISS and other spacecrafts. Molds also have the prospect to produce compounds not found in space, which could be beneficial on future space missions!

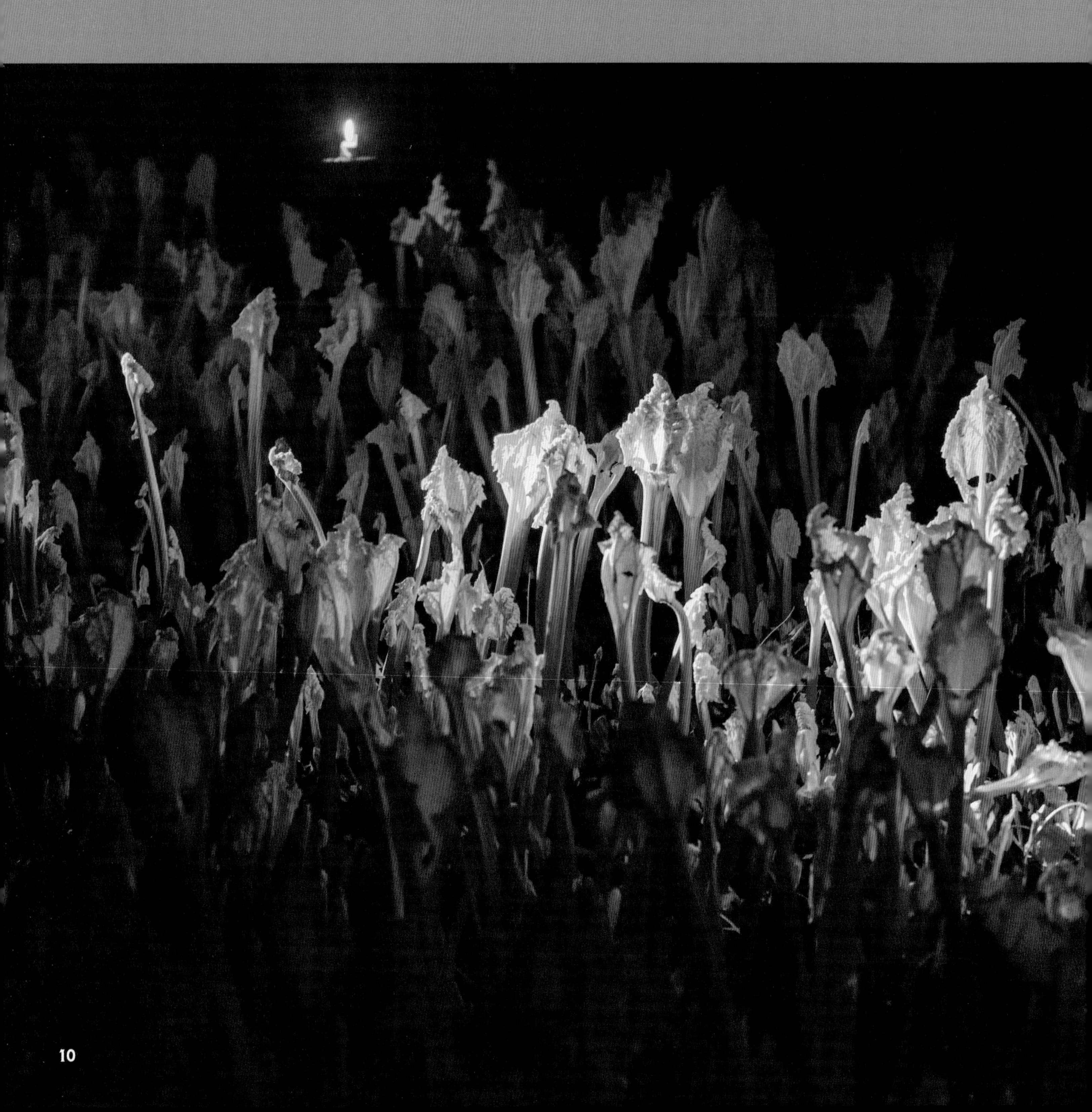

SINGING IN THE DARK

In 19th-century England, rhubarb was in such high demand that it exceeded the demand for opium. To keep up with the growing need for this vegetable, farmers developed a rather aggressive form of agriculture—forced growing.

The rhubarb is first left to grow in a field for two years, harvesting sunlight into their roots, and then they are placed in a forcing shed. This shed is heated and deprives them of all sunlight, which prevents the rhubarb from growing leaf blades. Without the need to grow leafage, the stalks can absorb all the nutrients, causing up to one inch of growth per day.

Though it sounds brutal, rhubarb grown in these conditions actually sing during the growing process! No, they don't sing along to folk songs. The rapid budding creates rhythmic cracking and popping noises.

Rhubarb is still farmed this way in places today, grown and harvested in candlelight. Believe it or not, forced-grown rhubarb is actually sweeter than free-range rhubarb!

CHEMISTRY

TIP FOR A TIPPLE

For those of you who are of legal drinking age, we know how to enhance your nip of whiskey—just add water!

Whiskey enthusiasts and connoisseurs may have told you that adding a few drops of water to your glass heightens the smell and taste, but scientists now have an explanation. The molecules in this liquid contain hydrophobic and hydrophilic parts, meaning half are repelled by water, while the other half are attracted to it.

The more water added to the whiskey, the more intense the flavor. During bottling, the average alcohol content of whiskey is about 45 percent, but some connoisseurs have been known to dilute their beverage to just 20 percent alcohol content!

GRAPE BALLS OF FIRE!

"Caution: Flammable" is what your package of grapes from the market should read. If you place a grape that has been cut slightly—but not all the way through—into the microwave, you'd better take a step back. A single zapped grape will shoot off a burst of plasma—the same stuff that makes up our sun!

Researchers have discovered the reason why these handheld snacks turn into miniature suns in the microwave—their size! Electromagnetic hotspots must be formed for plasma creation, but it takes two hotspots in contact to make plasma sparks. When the hotspots in each half of the grape reach the correct temperature, the sparks ionize sodium and potassium to produce plasma.

Some experiments have resulted in exploding microwaves, so don't try this at home!

ENVIRONMENT

UNDERWATER WATER

From land, the Angelita Cenote in Mexico's Yucatan Peninsula looks like an ordinary swimming hole. It's not until you dive almost 100 ft (30.5 m) that a second, hidden world appears.

A cenote is a deep sinkhole formed from the collapse of limestone bedrock that exposes groundwater underneath. Over time, cenotes can become filled with fresh rainwater. An underwater river is formed when the fresh top water meets the exposed salty groundwater.

The point where the two waters meet and cause a fog-like effect is called halocline. The different density levels of the two waters cause them to separate into layers. The result is a breathtaking convergence of two habitats that can make divers appear as though they are flying above an otherworldly river.

NO PAIN, NO PAIN

There are people who live among us born with what appears to be a superpower—the inability to feel pain.

One such person is Jo Cameron of Scotland. For the first 65 years of her life, she did not realize she was different, despite the countless bruises, burns, and broken bones she has experienced. She only became aware of her condition when she sought treatment for two notoriously painful ailments. To her doctors' surprise, however, she didn't report much discomfort and had no need for any painkillers. It was also noticed that her injuries tended to heal unusually quickly.

But the inability to feel pain is not always a good thing; people with similar conditions often find themselves injured without realizing it—sometimes very seriously. Dr. Ingo Kurth of the Institute of Human Genetics in Aachen, Germany, explains, "Pain is incredibly important to the process of learning how to modulate your physical activity without doing damage to your bodies, and in determining how much risk you take."

Geneticists took a close look at Cameron's DNA and found notable mutations on genes associated with pain sensation, mood, and memory. This unusual gene tweak also helps her feel shiningly optimistic and worry-free. By studying Cameron and people like her, scientists and doctors hope to find information that could help people who suffer from conditions like chronic pain, anxiety, and PTSD.

Believe It or Not! **. . . The brain processes pain but cannot feel pain itself!**

STUCK ON YOU(TUBE)

One of the latest reports from Hootsuite, a social media management platform, and digital marketing agency We Are Social has revealed that the world spends nearly 7 hours online each day.

The Philippines clocks up the most time online each day, at 10 hours and 2 minutes. Japan came in last, at 3 hours and 45 minutes. The United States and United Kingdom are closer to the global average (6 hours and 42 minutes). So what are people spending all that time on? The top five most visited sites were Google, YouTube, Facebook, Baidu, and Wikipedia. Good for you for opening a book.

RADIUM GIRLS

Glow-in-the-dark dial watches made popular in the 1940s clutter antique shops and family jewelry boxes. But these timepieces are more dangerous than they seem. The glowing effect was achieved by the use of radium-based paint applied by Radium Girls.

In the 1940s, working women would hand-paint watches by wetting and shaping the brush tips in their mouths and lips between dipping the brush into the paint. At the time, radium was considered a health and beauty product, and was believed to be safe and even beneficial to the body.

Over time, radium emits radon in the form of an odorless and colorless gas, which causes rapid decay and can lead to cancer. The health of Radium Girls quickly declined, and the most common problem was rapid tooth loss, accompanied by unhealable gum ulcers. The jawbones of some young women eventually crumbled away, and many died young from intense internal bleeding.

The Radium Girls took their illnesses to court, challenging that their employers were responsible for not protecting them against harmful side effects. Ultimately, they won their case, and awareness of workplace safety paved the way for the Occupational Safety and Health Administration (OSHA) to be established in the 1970s.

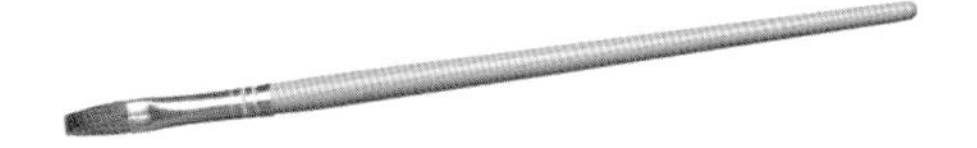

CHEMISTRY

FIZZY FEELINGS

Every time you pop open a can of soda and enjoy the fizzing sensation, you can thank your taste buds! You may believe the sensation you're experiencing is simply the feeling of bursting bubbles, but you're actually tasting the carbonation.

The same cells on your tongue used to sense sour stimuli are responsible for detecting the fizzy carbon dioxide. In a research study conducted using a pressurized chamber that removed the fizziness of the soda, people said there was no difference in taste whatsoever, meaning the flavor of the soda pop doesn't come from the bubbles—you're actually tasting the carbon dioxide!

Believe It or Not! . . . **Adult cats don't meow to each other—just to humans!**

A-MEOW-ZING MOVEMENTS

It is often said that cats always land on their feet, but there's actually scientific reasoning behind this phenomenon. No, it's not that cats have nine lives—which we still have no evidence of. It's due to something known as the "righting reflex!"

Including their tails, felines have between 48–53 vertebrae, while humans have only 33. Their flexible backbones, combined with their highly tuned sense of balance, allow them to twist midair and orient their bodies.

When cats jump or fall from a high place, they utilize an inner-ear balance system called the vestibular apparatus to determine up from down. Once cats rotate their upper body to have their paws face down, their lower body follows suit, allowing them to land safely on all of their toe beans.

Believe it or not, kittens as young as seven weeks can master this skill!

THE LIFE AND TIMES OF PHINEAS GAGE

In September 1848, 25-year-old Phineas Gage survived being blasted through the skull with a 3.5-ft-long (1-m), 1.25-in-thick (3.175-cm) iron rod and cemented himself in the annals of medical history as one of the first cases to show a link between personality change and brain trauma.

Gage was a foreman working for a railroad company in Vermont, blasting rocks to clear space for tracks to be laid. Holes would be drilled into rocks and then filled with gunpowder and topped with sand, which was then packed tight with a tamping iron. On this fateful day, Gage went to pack the powder without realizing the sand hadn't been added; the iron ignited the powder and sent the rod through Gage's face and out the top of his skull.

Astonishingly, he survived. It's reported he didn't lose consciousness, spat out "about half a teacupful" of his brain matter, and told the attending physician, "Here is business enough for you." Months later he made a seemingly full recovery, but the once model worker and mild-mannered man had become disrespectful and hotheaded.

It is often said that his personality change was permanent, but doctors he visited years later made no record of his outward behavior. After years of moving around (farming in New Hampshire, making appearances as a medical oddity at P. T. Barnum's American Museum in New York City, and working on a stagecoach in Chile), Gage began suffering seizures and died at the age of 37—12 years after his brain injury. His skull and the offending tamping iron can be seen today at the Warren Anatomical Museum near Boston.

WHY YOU CAN'T STAND YOUR OWN VOICE

Why is it that the voice you hear played back on a speaker doesn't sound like the voice you've heard coming out of your mouth for all these years?

We hear sounds by vibrations being picked up by our eardrum. We perceive external sounds, like a beeping car or a radio, through sound waves passing through the air into our ear canals, into our inner ear, and on to our cochlea. When our voice is played back to us via a speaker, we are hearing air-conducted vibrations.

A lot of what we hear when we speak is perceived in the same way as external noise, but we also pick up on vibrations that have come through our jawbone and skull. This is known as inertial bone conduction, which tends to "bring out" the lower-frequency vibrations, making your voice sound deeper and less squeaky than it actually is. In all likelihood, the fact you don't like the sound of it is simply because you are not used to it.

Unfortunately, the depressing reality is that the awful noise you hear when you play back a recording of your voice is actually how your voice sounds to the 7.6 billion other humans on Earth. Sorry about that, folks.

BETWEEN TWO CONTINENTS

Have you ever wanted to be in two places at once? How about two separate continents? Believe it or not, it's possible within Iceland's Thingvellir National Park. And while this version of being in two places at once won't make you any more productive, it does make for an unbeatable experience and photo op.

The Silfra Crack is a narrow, water-filled chasm that happens to be part of the Mid-Atlantic Ridge, where the North American and the Eurasian tectonic plates are diverging. Most of the Mid-Atlantic Ridge is, as the name suggests, in the middle of the Atlantic Ocean floor. But Iceland is one of the few places, and the largest, where portions of it are above the water.

The volcanic activity of the ridge has created multiple fissures across the country. Over time, glaciers have melted, been filtered through miles of porous volcanic rock, and filled in the Silfra Crack, creating an underwater wonderland where one can float between the continental plates. The water is considered some of the clearest and calmest in the world; divers compare swimming in it to floating in space.

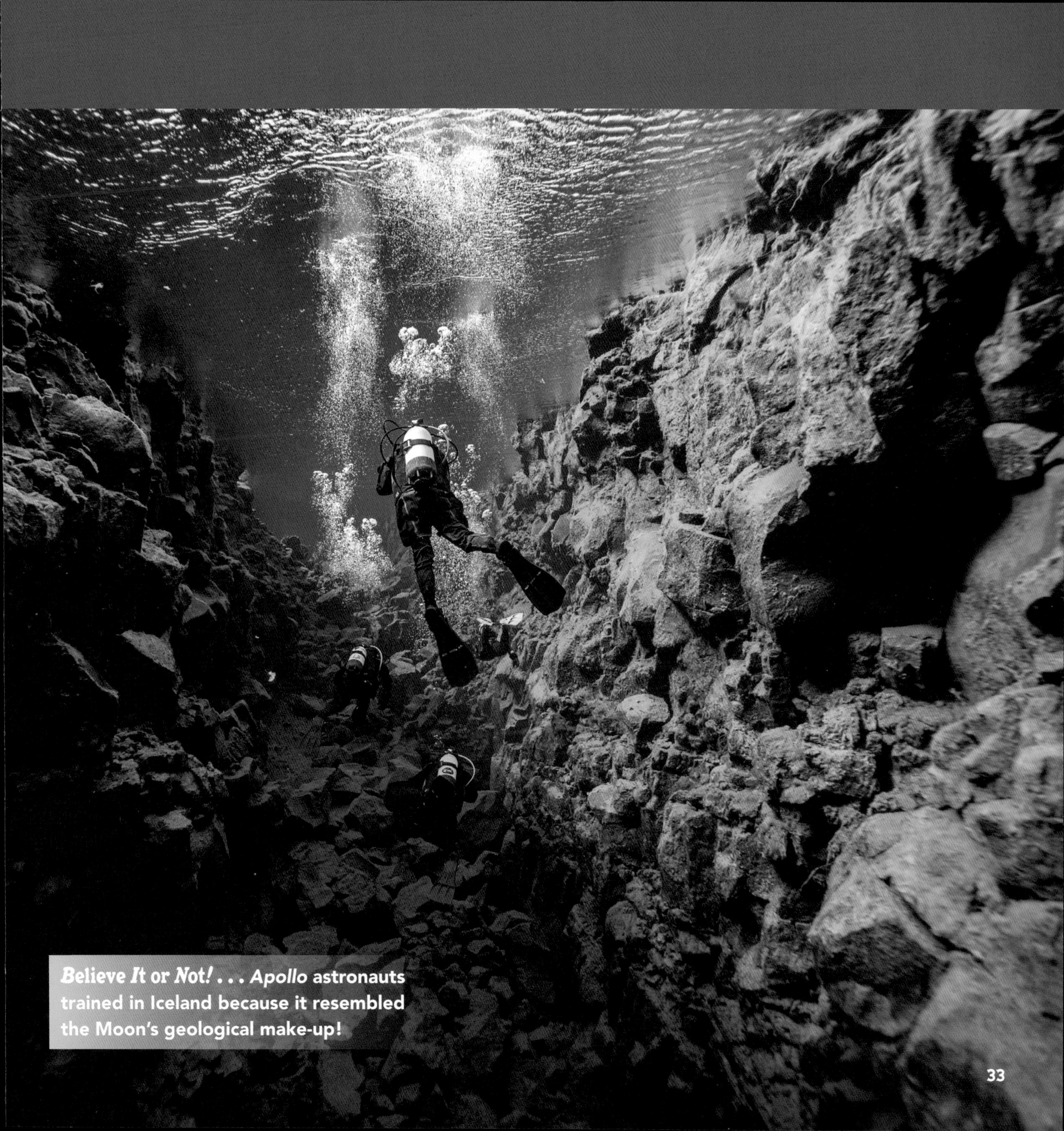

Believe It* or *Not! **. . . *Apollo* astronauts trained in Iceland because it resembled the Moon's geological make-up!**

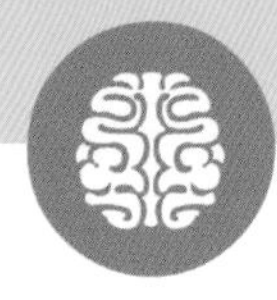

KNOCKING ON DEATH'S DOOR

Trying to find out what happens to our minds when we die is one of life's biggest conundrums. Researchers can't predict when someone will suddenly die, and they can't ethically bring someone to the edge of death in a controlled experiment. That's why the phenomenon of Near Death Experiences (NDEs) is one of the most fascinating elements of the natural—or supernatural—world.

NDEs are rare and unusual moments that occur when someone is perilously close to the brink of death. They are typically recounted after recovery and are described as an out-of-body experience with a bright tunnel of light.

A study published in the journal *Mindfulness* in December 2018 documented 12 Buddhist monks who are highly proficient in the art of self-meditation and put themselves into a near-death state. During these NDE meditation sessions, the monks reported several phases their mind experienced. These included losing the sense of their bodies, losing the concept of time, visualizing otherworldly beings, and feeling a sense of emptiness.

Though these monks reported the ability to control their mind during these NDEs, we highly recommend you don't try this at home.

ROYAL BLUE

In 1810, King George III slipped into pure madness and was soon forced to retire from public life completely. He may be best known in the history books for being the king who lost America, but did you know he had blue urine? Could his mental illness and streams of blue be related? It could all come down to a genetic blood disorder called porphyria.

In the last decade of his life, King George III had lost a majority of his sight and was in constant pain. His once worldly and rich vocabulary quickly diminished as he began constantly repeating himself and writing long, confusing letters. It is also rumored he would walk around completely nude. On top of it all, the blue urine…

A 1969 study suggested King George III suffered from porphyria, a disorder caused by an over-accumulation of porphyrin, which helps hemoglobin, the protein that moves oxygen throughout the body. In some cases, porphyrin is excreted in the urine, giving it a purple or blue hue.

Acute porphyria can also seriously affect the nervous system. Symptoms include hallucinations, delirium, insomnia, anxiety, and even paranoia. Moreover, the king's doctors might have worsened this condition and its symptoms by treating George with doses of arsenic, basically poisoning him.

LOOKING FOR LIFE

In the 1960s, Dr. Frank Drake came up with an equation that tries to assess the likelihood of life starting on a planet. The equation includes factors like the fraction of formed stars that have planets and the average number of those planets in the habitable zone. Even when incredibly conservative estimates are used, the Drake equation suggests the universe is teeming with life—so where is it?

This is a problem that has long plagued astrophysicists, and there's no clear answer. It is known as the Fermi Paradox, and there are a number of possible solutions—some more unnerving than others.

It's possible that space is just too big or we haven't looked hard enough. Another solution suggests there is a Great Filter in the universe, at which intelligent life stops, perhaps through self-destruction or for other reasons. It could be that we are the first species to pass this filter, or that we are yet to reach it—and all other intelligent civilizations before us have been destroyed. Eek.

There is another answer to the Fermi Paradox: perhaps we are alone in this universe. As the late Sir Arthur C. Clarke once famously said, "Two possibilities exist: either we are alone in the universe or we are not. Both are equally terrifying."

TINY BUT TOUGH

Lions, tigers, and water bears! Oh my!

The tardigrade, also known as a water bear, a moss piglet, or a slow walker, is a tiny invertebrate that can inhabit almost any place on Earth. They measure around half a millimeter and have four little pairs of legs with claws on each end. These claws give them the appearance of a bear, lending them the nickname *water bear*!

Tardigrades are extremophiles, meaning they can survive in the most extreme living conditions most other organisms wouldn't have a chance in. From freshwater to salt water to the moist moss in your backyard, water bears can thrive in temperatures ranging from near absolute zero (−460°F/−273°C) to a balmy 248°F (120°C)!

These miniscule creatures have the ability to enter a stage called cryptobiosis, where they dry up, curl up into a ball, create a hard outer shell, and stay like that for more than 100 years. They rehydrate themselves when living conditions improve. Can you say nap royalty?

BONE MUSIC

During a time when plastic vinyl was hard to come by and music was heavily censored, Soviet-era Russians would share and play tunes off of exposed X-ray film.

Before CDs, MP3s, and streaming, vinyl records were once the most popular way to enjoy music (and have made quite the comeback). To play music, vinyl records are covered in thin grooves that are translated into sound by delicately placing a needle into the channels and then spinning the record on a turntable.

The X-ray records work in the same way, but instead of using the black plastic of typical vinyl records, *stilyagi* used exposed X-ray film scavenged from hospital dumpsters. The *stilyagi* were a Soviet-era subculture of young Russians fascinated by Western pop culture, which was heavily censored by the government. Illegal copies of songs were etched into the X-ray film, which would then be cut into a circle and then burned with a cigarette in the middle so the record could be placed on a turntable. And no matter what genre of music was copied onto the makeshift record, that's pretty punk.

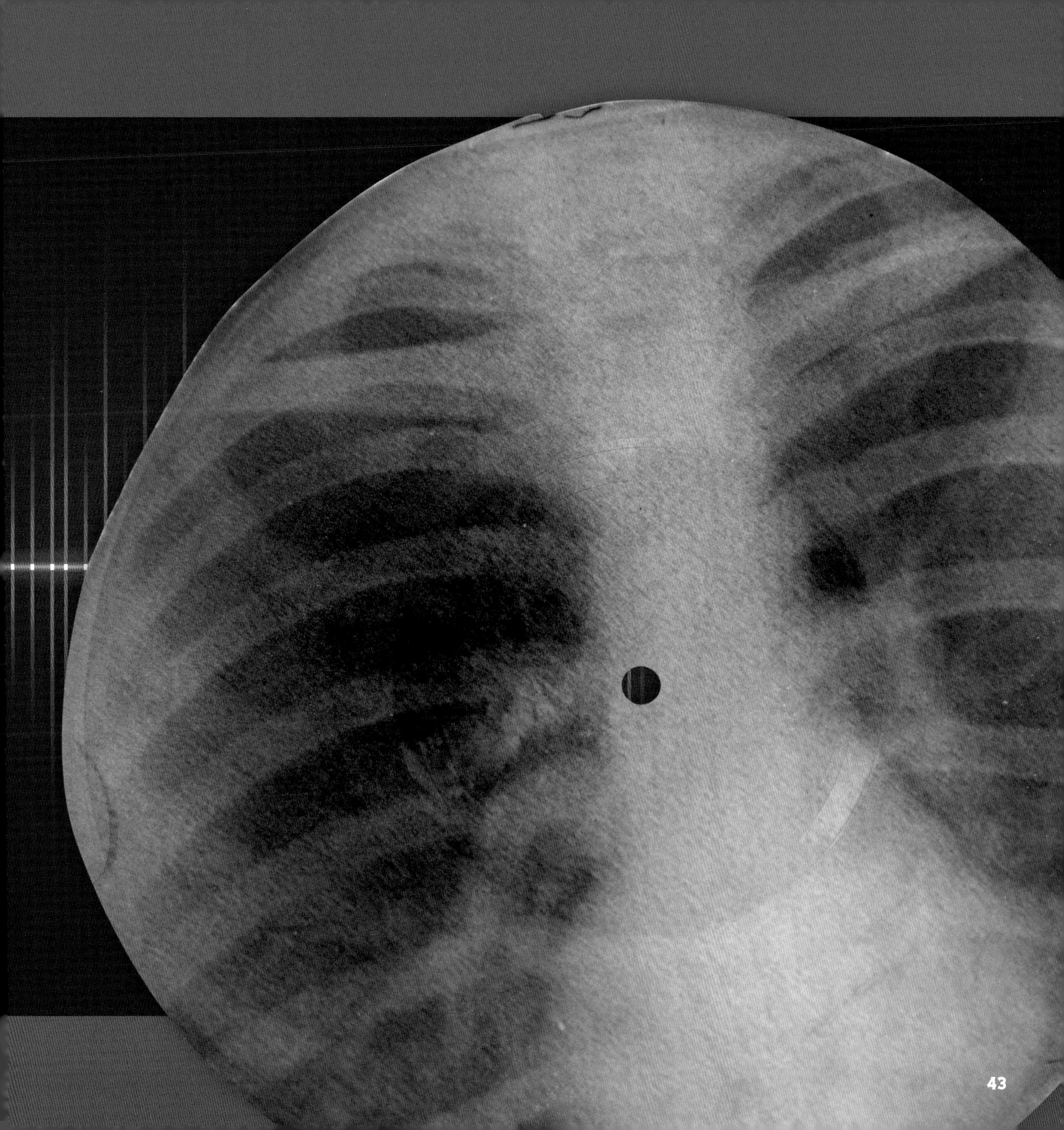

Believe It* or *Not! . . . An onion has a genetic code five times longer than that of a human being!

FIGHTING BACK THE TEARS

Why is it that onions make you cry when you begin to mercilessly dissect them? It's fairly unlikely that it's because you've formed an emotional bond with your onion and are sad to see it chopped up into little pieces—so it must boil down to some rather curious chemistry.

Onions belong to a group of vegetables that absorb high quantities of sulfur compounds from the soil. These are converted by the onion plant into other compounds that can readily transmogrify into a gas.

Murdering an onion splits open cells that release enzymes and these gas-prone compounds—combining the two forms into a jarringly named gas, syn-propanethial s-oxide, which quickly reaches your eyes. Detecting this chemical irritant, your eyes send a signal to your central nervous system, which in turn causes your eyes to weep in an attempt to wash it out.

Unfortunately, it wouldn't be a good idea to genetically modify onions to remove these pesky compounds: They're actually partly responsible for making onions so tasty in the first place.

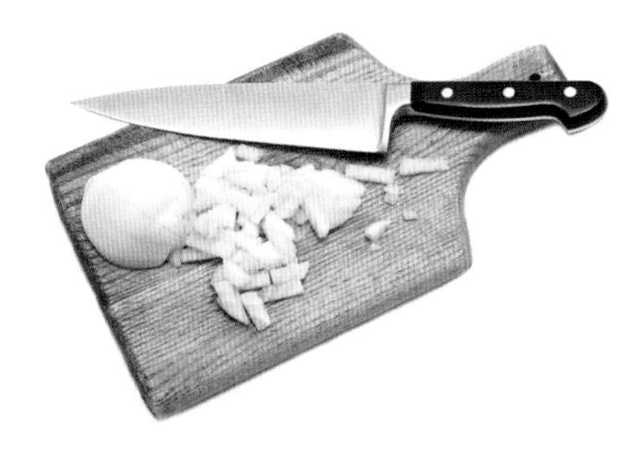

Believe It** or **Not! . . . A raisin will bounce up and down continuously when dropped in a glass of champagne!

PHYSICS

RAISE A GLASS

A popping champagne cork produces freezing jets of carbon dioxide and, unexpectedly, shock waves like the ones released by fighter jets.

Before a bottle of champagne has been opened, a mixture of pressurized carbon dioxide and water in gaseous form lies trapped beneath the cork in the bottle's "headspace," or neck. With the use of high-speed cameras, a group of researchers have found that when the cork flies off, that trapped gas is released, cooling down and condensing in the process, which forms a jet of dry ice that whooshes out of the bottle faster than the speed of sound.

Meanwhile, the process creates something known as a Mach disk, the kind of shock wave you'd see in the exhaust streams whizzing out of rockets and jets. The Mach disk exits the bottle in the plume of CO_2 and water vapor, vanishing in a mere millisecond. The researchers noted that it is safer to uncork a bottle gently with a "subdued sigh" to avoid eye-related incidents, but concede that a dramatic pop is more fun.

SALTY REFLECTIONS

Stretching for 4,086 sq mi (10,582 sq km) across the country of Bolivia, Salar de Uyuni is the world's largest mirror—eight times the size of New York City!

About 30,000 to 40,000 years ago, a giant prehistoric lake called Lake Minchin went through a series of transformations alongside several other vast lakes. When it dried, it left behind two bodies of water and major salt deserts. During the rainy season, the salt flat is covered with a thin sheet of water due to the floods and overflow from the neighboring lakes. It reflects light, creating a mirror effect under the vast open sky. Many people refer to it as "the place where Heaven meets Earth."

Salar de Uyuni sits 11,995 ft (3,656 m) above sea level, making it easily visible from space due to its bright white color and enormous size.

Believe It or Not! **... NASA and other space agencies sometimes use this salt flat's surface to calibrate satellite orbits.**

TALK TO THE HAND

Stringing together letters or symbols to create a coherent thought is part of the communication process. The ability to understand one another's thoughts is fundamental to a shared language.

But what if you were never exposed to language? Could you be taught something you never knew existed? In 1970, Susan Schaller, an American Sign Language interpreter in Los Angeles, California, did just that.

While interpreting at a community college, Schaller met Ildefonso, a 27-year-old student who was born deaf. After introducing herself, instead of signing back to Schaller with his own name, he copied her movements. He recognized her movements as commands to be mimicked instead of representations of abstract concepts, such as a wave to mean "hello."

Because Ildefonso thought every symbol was commanding him to do something, teaching him basic sign language was out of the question. So Schaller ignored him. During their sessions, she taught to an invisible student in an empty chair while Ildefonso watched from the sidelines. Schaller role-played as both the student and the teacher, hopping into the chair to mimic a student understanding the language.

After weeks of lessons, something clicked inside Ildefonso's mind as he realized every object, person, and movement has a name. At 27 years old, Ildefonso learned language.

STRIPES? ON. MOSQUITOES? OFF!

It might not be the most dignified look in the world, but a new study has found it might be worth painting your cows to look like zebras.

Experiments have shown that horseflies tend to avoid black-and-white striped surfaces, while other studies have suggested that the stripes may cause a kind of motion camouflage targeted at the insects' vision, confusing them much in the way that optical illusions confuse us.

To find out whether this can be applied to cows, Japanese researchers painted white stripes on a group of black cows and black stripes (on black bodies) on another, leaving another group of cows unpainted as control moos.

The cows were then observed for fly-repelling behaviors (head throws, ear beats, leg stamps, skin twitches, and tail flicks), and the number of flies landing on their bodies was counted. The zebra-cows were found to have more than 50 percent fewer biting flies on their bodies and a 20 percent decrease in fly-repelling behaviors versus the black-striped cows and the control group.

The researchers posit that painting stripes on livestock would not only be great for cow health but could also help reduce the need for pesticides. Sounds like a win-win (and a fashionable one at that).

Believe It or Not! **. . . Some species of mosquitoes fly up to 40 mi (64 km) to find a meal of blood!**

UPSET STOMACH

On June 6, 1822, 28-year-old Canadian fur trapper Alexis St. Martin was accidentally shot at short range in the stomach by a duck hunter, an injury that would have left most people of the time dead as a doornail. He lucked out and survived, but at a cost.

Dr. William Beaumont was able to save St. Martin, but he ended up with a gaping 2.4-in (6-cm) hole, or "gastric fistula," in his side when the hole in his stomach fused to the opening in his abdomen as he healed. But it wasn't all bad news, as Beaumont was able to observe digestion in real time by tying a silk string to pieces of beef, chicken, bread, and cabbage, and dangling them through the hole into St. Martin's stomach.

St. Martin's stint as a medical guinea pig lasted for about three decades, as he frequently returned to Dr. Beaumont for experiments in exchange for money and the promise that the hole would be sealed. The latter never happened, and the men parted ways when St. Martin asked for more money than Beaumont was willing to pay.

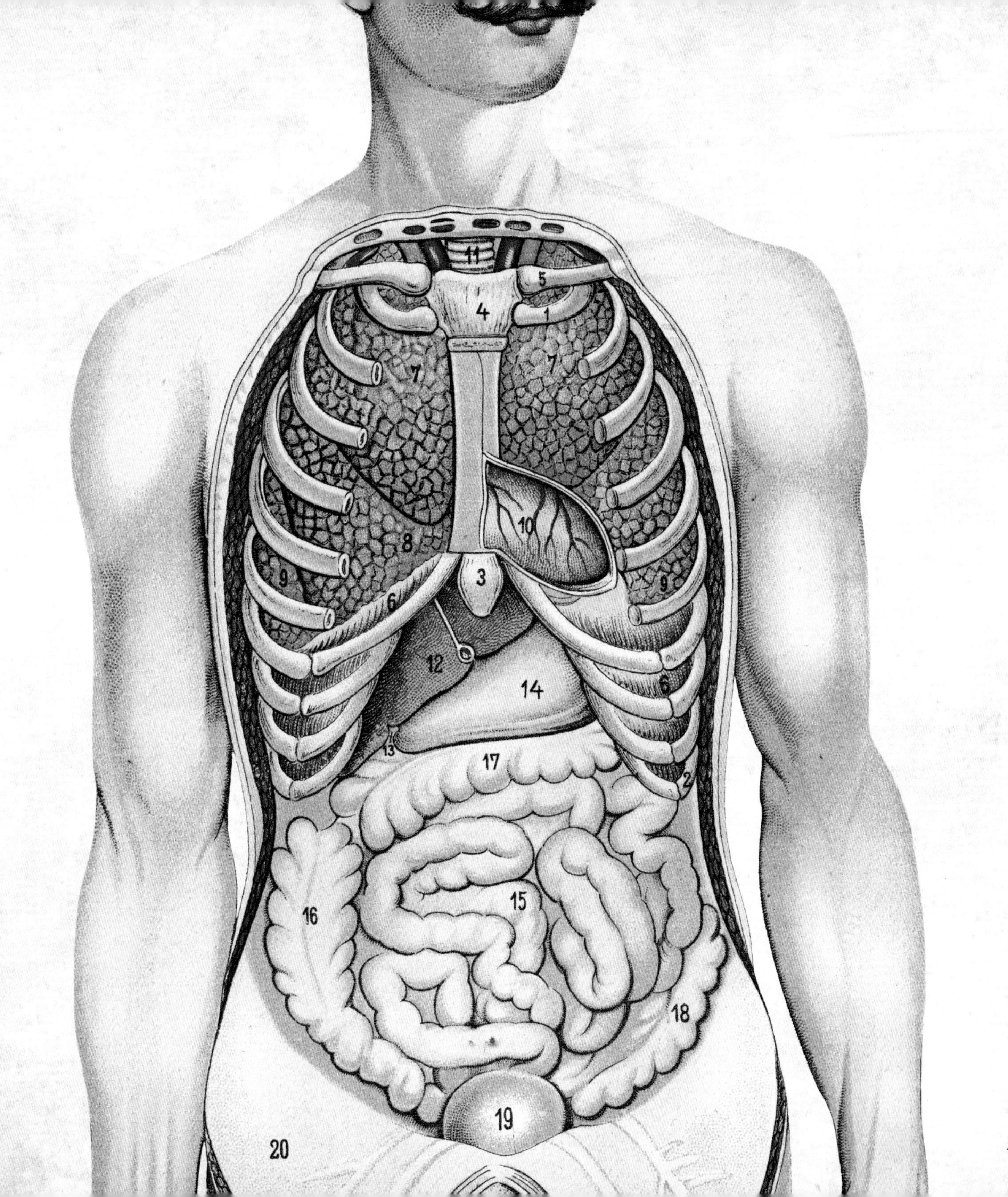
11
5
4
1
7
7
10
8
3
9
9
6
12
14
6
13
17
2
15
16
18
19
20

UNITED STATES OF AMERICA

THE UNITED STATES' DOOMSDAY PLANE

Should nuclear war break out, one of the safest places to shield yourself from the fallout is aboard a plane. More specifically, the U.S. Air Force's E-4B, or the Doomsday Plane.

Used to shuttle the Secretary of Defense from place to place, the E-4B is a modified Boeing 747 that comes in at nearly six stories tall. It boasts 18 bunks, six bathrooms, a battle staff work area, executive quarters, enough space for a 112-strong crew, and more. It can stay in flight for days at a time and refuel midair. There are no windows or digital touch screens. Instead, analog technology is used because it is less vulnerable to electromagnetic pulses. On top is a "ray dome" that houses about 60 satellite dishes and antennas that can be used to communicate with ships, subs, aircraft, and landlines around the world. And that's just what the public knows—most of its capabilities are classified.

As safe as E-4Bs may be against man-made disaster, apparently they are no match for mother nature. In 2017, a tornado struck the Offutt Air Force Base in Nebraska, leaving two of the four flying behemoths nonoperational for months.

LYRE LIAR

The Australian lyrebird can mimic the songs of as many as 25 other bird species, as well as imitate animal sounds and a whole range of human noises, including chainsaws, car engines, trains, car alarms, fire alarms, rifle shots, camera shutters, human voices, crying babies, and cell phone ringtones.

The superb lyrebird's ability to mimic almost any sound it hears is thought to stem from the makeup of its vocal organ, or syrinx. Whereas other songbirds have four pairs of syringeal muscles, the lyrebird only has three, making it more flexible.

Nobody really knows why the lyrebird needs such a vast repertoire. Some experts believe it uses its song simply to intimidate rivals in the forest. And believe it or not, the lyrebird's name does not come from its musical abilities, but rather the shape of the male's tail when spread in its courtship display.

THE SHOCKING TRUTH

If you have ever rubbed a balloon on your head and watched your hair stand on end, received a small shock from touching metal after rubbing socked feet along the carpet, or seen small sparks when brushing fur in the dark, then you've experienced static electricity!

We know that rubbing two objects together creates static electricity, but scientists have finally uncovered the reason why it is created. Even on a microscopic level, no material is perfectly smooth, and parts of the surface protrude. These protrusions are bent by the force of rubbing them against another objects, which results in the "flexoelectric effect," or the charge that emits from them under stress. Shocking.

STAINING SPECIMENS

People who prepare animal specimens for museums and schools—as well as for personal collections—can make snakes, frogs, mice, and other small animals into scientific works of art called "diaphonized specimens." This technique, also known as "clearing and staining," helps people understand the internal anatomy of an animal in a unique way.

The process is complicated, involves somewhat dangerous chemicals, and takes a lot of training to perfect. In one step of the lengthy process, a preparator must soak the dead animal in a chemical called trypsin, which destroys a color-creating protein called casein. No casein, no color. After this part of the process, only a transparent version of the original animal remains.

Next, the preparator colors the specimen with types of dye that are attracted to certain substances in the body. As a result, depending on what dyes are used, collagen-rich bones usually turn red or pink, and the remaining cartilage can turn blue. The result is a colorful, mystical-looking specimen that is both artistic and educational.

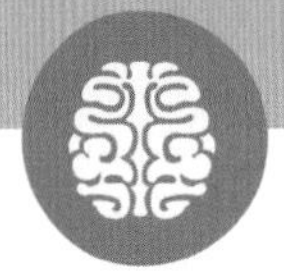

WAKING NIGHTMARE

Nightmares and dreams occur when the body is deep in REM (Rapid Eye Movement) sleep, where muscle functions are turned off. But if you are unable to move your body when falling asleep or just waking up, that phenomenon is called sleep paralysis.

Considered a parasomnia, an episode of paralysis is involuntary and occurs when the brain regains consciousness prior to your body catching on. These episodes can last anywhere from several seconds to several minutes. Paralysis usually ends on its own, but can also be interrupted by making an intense effort to move or by someone touching or speaking to you.

Sleep paralysis can inhibit your ability to speak or move your hands, arms, feet, legs, and head. Your breathing functions are normal and you are fully aware of your surroundings, but that doesn't make it any less terrifying. Common symptoms are feelings of anxiety and fear, but many cases describe hallucinations of hearing or seeing things that aren't real or the feeling of another person standing in the room.

It is possible that this 1781 painting, *The Nightmare* by Henry Fuseli, is an interpretation of sleep paralysis. Sufferers often report feeling a heavy weight or seeing evil creatures sitting on their chests.

Believe It or Not! **. . . Sleep paralysis is a common explanation for people who claim to have been abducted by aliens in their sleep.**

BETTER OUT THAN IN

While humans just vomit the contents of their stomach up the way they came in, a few animals vomit their entire stomach instead of hurling. Known as stomach eversion, it's a thorough—if inelegant—way to clean your insides.

Many species of frogs will vomit out their own stomach. Some can even be seen giving their stomach a quick scrub with their feet. After pulling any remnant food off of their stomach, they swallow their organs, as if nothing happened at all. This process is normally incredibly quick, taking less than half a second.

Sharks also exhibit this stomach-churning tactic. Bones, feathers, turtle shells—anything that a shark should not have consumed—must also be ejected this way. Additionally, if a shark feels it is in danger, it will react by vomiting the contents of its stomach. With an empty belly, the shark can possibly swim more quickly to safety.

Believe it or not, some gluttonous sharks will vomit simply so they can eat more. Sharks feeding on baleen whale carcasses are known to vomit once full and then immediately return to eating. Maybe that's why they've been banned from buffets.

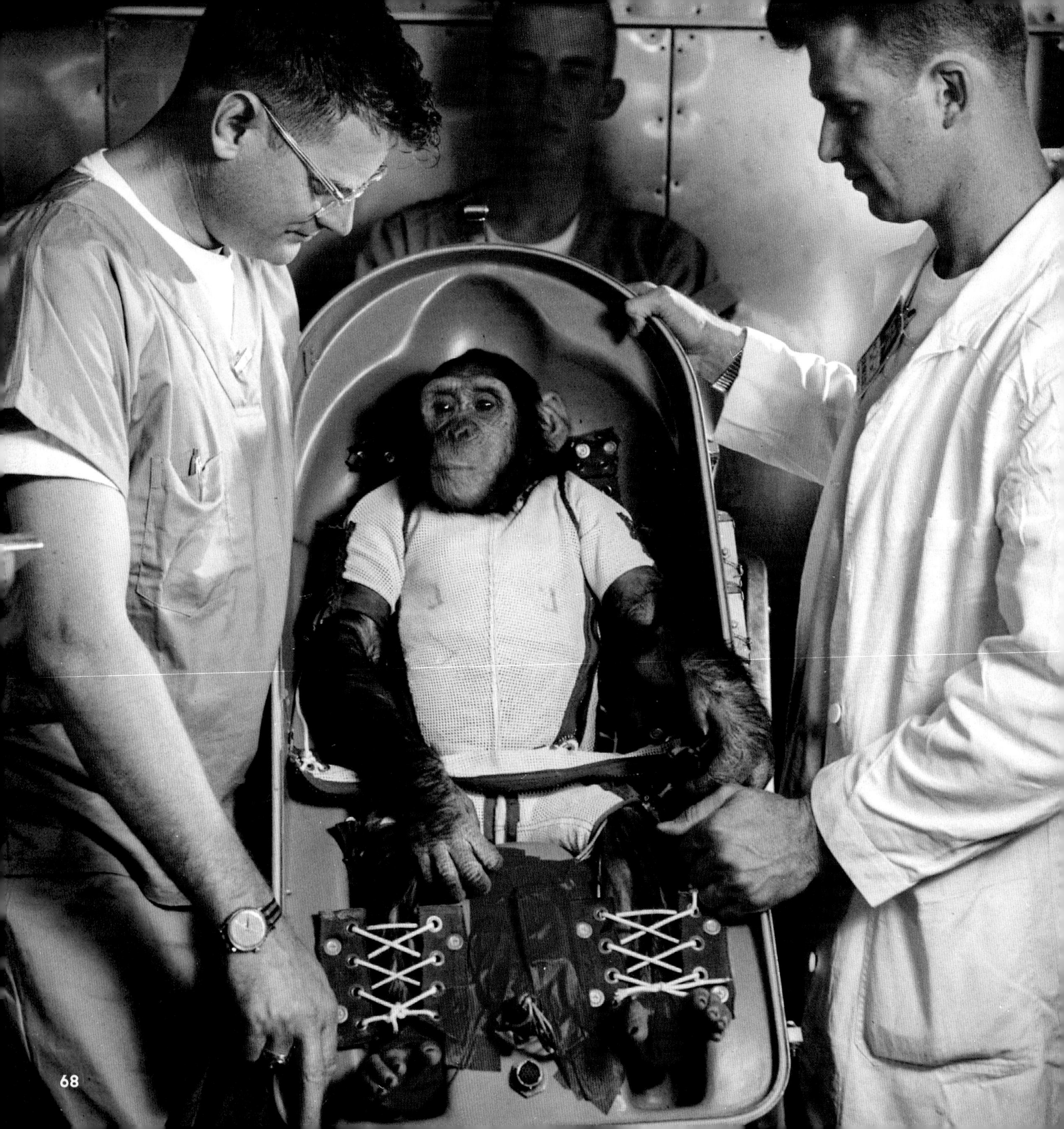

HAM THE SPACE CHIMP

Three months before Alan Shepard became the first American human in space, the United States launched a chimpanzee astronaut. His name was Ham.

For two years, number 65, as he was known (officials were worried that bad publicity might result from the death of a named chimp), was given intensive training at Holloman Air Force Base, New Mexico. He was taught to push a lever within five seconds of seeing a flashing blue light.

On January 31, 1961, the chimp, dressed in a mini space suit, was launched aboard a Mercury-Redstone rocket from Cape Canaveral, Florida. During his suborbital flight, computers on the ground measured normal vital signs, letting mission control know their brave chimp was alive.

He performed his tasks admirably, and his capsule touched down safely in the Atlantic at the end of the 16-minute flight. Though he pulled his lever just slightly slower in space than he did on Earth, this feat proved that human motor control was possible in space. Only when he had safely returned to Earth with nothing worse than a bruised nose was he renamed Ham.

VIRTUALLY UNWRAPPED

In 79 CE, as Mount Vesuvius rained down hell on the towns of Pompeii and Herculaneum below, a fine set of scrolls laid in a private library near the coastline. Along with the towns and their people, the scrolls were carbonized through a blast of hot volcanic debris, searing them into lumps of brittle carbon that are too fragile to unravel.

Collectively known as the Herculaneum papyri, the texts are thought to be the only surviving library from antiquity that exists in its entirety. Now, almost 2,000 years later, a team of researchers say they finally have the technology to decipher the papyrus text.

Scientists from the University of Kentucky have employed the help of high-energy X-rays to pick up on subtle hints of ink that are invisible to the naked eye. They will then use artificial intelligence to "fill in the gaps." The texts that have been successfully studied namely contain writings of a philosophical nature that provide insight into the world of the Roman Empire.

Professor Brent Seales, director of the Digital Restoration Initiative at the University of Kentucky.

BIKINI ATOLL IS THE BOMB

Bikini Atoll might sound like an idyllic island getaway, but any visitors would be better off wearing a biohazard suit than swimwear, as the nuclear fallout from tests carried out by the U.S. military more than 60 years ago has left parts of the Marshall Islands more radioactive than the infamous Chernobyl.

Between 1946 and 1958, the United States conducted almost 70 nuclear bomb tests in the Marshall Islands, a chain of atolls and volcanic islands in the central Pacific Ocean. The largest of these detonations released the equivalent of 15 megatons of TNT in Bikini Atoll on March 1, 1954. The explosion completely vaporized a nearby artificial island and left a crater measuring 1.2 mi (2 km) in diameter and 250 ft (76 m) in depth. Now that's a spicy meatball.

Believe It or Not! **. . . Goosebumps are caused by tiny muscles attached to each and every hair follicle on your skin!**

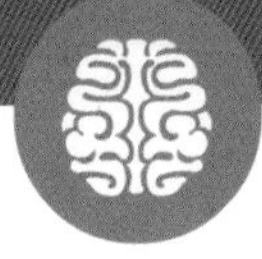

CAREFUL WHISPERS

It might seem like just another weird corner of the internet, but ASMR, short for Autonomous Sensory Meridian Response, has become incredibly popular on sites like YouTube. On these platforms, ASMRtists aim to trigger ASM responses in listeners by doing things like speaking softly; making quiet, repetitive sounds (like turning pages in a book); and performing mundane tasks.

Only some people appear to experience ASMR, and they describe it as a pleasant tingling that starts at the crown of the head and runs down the neck, usually sparked by watching or listening to certain stimuli.

It's not a new thing, either. The internet has made it easier to create a virtual community around ASMR, but some believe there is a reference to the sensation in Virginia Woolf's 1925 novel *Mrs. Dalloway*: "… with a roughness in her voice like a grasshopper's, which rasped his spine deliciously and sent running up into his brain waves of sound which, concussing, broke."

Scientists have recently looked at the psychological underpinnings of these "brain tingles," and it turns out they are actually pretty good for you. Along with decreasing levels of stress and sadness, ASMR was even shown to decrease people's heart rates.

MOLTEN MELTDOWN

When the roof of reactor 4 at the VI Lenin Nuclear Power Plant in Ukraine blew off in an uncontrolled explosion on April 26, 1986, around 57 metric tons of uranium were ejected into the atmosphere. The effects of such intense radiation displaced more than 300,000 people within the 18.6-mi (30-km) stretch of the Chernobyl Exclusion Zone. At least 237 people suffered acute radiation sickness, and the World Health Organization expects 4,000 have died or will die due to radiation effects.

The area that received the most intense dose of radiation was the Red Forest. Pine trees died out quickly due to the hot nuclear fuel burning their leaves, but deciduous plants that could drop their leaves survived the blast. Within two years after the disaster, the mammal population steadily increased. The lack of human threat and hunting has allowed animal populations to flourish.

If you want to bring a little piece of Chernobyl home, scientists from the United Kingdom and Ukraine have developed Atomik Vodka, made from radioactive grain from the Chernobyl Exclusion Zone!

Believe It or Not! **. . . This cooled molten mass of radioactive material, known as "the Elephant's Foot," once flowed like lava and emitted deadly levels of ionizing radiation, which is what distorted this photo.**

Believe It or Not! **. . . Cheddar cheese contains more tryptophan than turkey!**

ASIAGO A DAY

Great news: cheese, mmm cheese, might actually protect you from some of the perils associated with eating too much salt.

While we all need to consume a bit of salt to keep our bodies ticking, too much can lead to high blood pressure, heart disease, strokes, and even dementia. In a world filled with fast food and processed meals, avoiding salt can be tricky, but new research suggests that eating dairy might help counteract its effects.

The results of a small study of 11 people with salt-sensitive blood pressure suggest "that people who consume the recommended number of dairy servings each day typically have lower blood pressure and better cardiovascular health in general," according to Lacy Alexander, Professor of Kinesiology at Penn State and co-author of the study.

However, before you start feasting on salty foods with a side of cheese thinking you're being healthy, there are a few caveats to take note of. First, no matter how much dairy you eat, it's best to keep your salt intake nice and low at the recommended 1,500 mg per day. And while cheese is both delicious and a great source of calcium, it can be high in saturated fat and salt. Everything in moderation—even fondue.

MIGHTY MONSTERS

Though a vast majority of the ocean floor has yet to be explored, this metallic and malicious creature may have you thinking twice about dragging your feet. *Eunice aphroditois*, commonly known as the bobbit worm, is a bristle worm that inhabits burrows it creates on the sea floor. The bobbit worm is only about 1 in (2.5 cm) wide but can grow anywhere from 4 in (10 cm) to 10 ft (3 m) long!

Once inside its sandy burrow, this ambush predator uses five antennae to sense moving prey and strikes with sharp, powerful mandibles—sometimes splitting a fish in half! The bobbit worm injects prey with a toxin, making it easier to digest. This bristle worm is typically found in warmer waters and hunts in a stationary position under soft silt or among coral reefs. Unfortunately for professional and hobby aquarists alike, bobbit worms have also been known to invade aquariums, hiding and surviving inside rocks collected from the ocean used to recreate the sea environment. The aquarists usually don't realize the worm is there until their fish start mysteriously disappearing.

OUR FILTHY HABITS

Online video streaming makes up 60 percent of the world's data traffic and generated more than 300 million tons of carbon dioxide during 2018—and nearly a third of it was porn.

Carbon emissions are produced by the electricity needed to power the consumption and production of digital equipment, including everything from data centers to your smartphone. According to a study by French think tank The Shift Project, porn makes up around 27 percent of the word's video streaming. That means it pumps out around 100 million tons of carbon dioxide each year—more than the total annual output of Belgium. Maybe next time just use your imagination, for the sake of the environment.

THROWING RED ROCKS IN GEL HOUSES

To avoid drastically altering Mars's atmosphere in order to make it livable for humans, a team of researchers instead propose encasing enormous areas of Mars in an actual greenhouse made from silica aerogel.

Aerogels are formed of interconnected nanoscale clusters surrounding pockets of gas. Being 97 percent porous, aerogel is exceptionally light but also an excellent insulator. It's already used to insulate the rovers from the dangerous cold of the Martian night.

Silica aerogel has several features that make it particularly suited for the task of colonizing Mars. It blocks ultraviolet light but lets visible wavelengths through, so it would allow enough radiation to reach the Martian surface for plants to grow. Meanwhile, it is so light that the shield could be built on an unthinkable scale. Laboratory testing suggests that when exposed to radiation matching Martian sunlight, just 1 in (2.5 cm) of aerogel can warm the area beneath by 150°F (65°C), close to the difference between Mars and Earth.

This approach would also avoid some of the ethical problems with terraforming Mars, potentially supporting millions of people while leaving most of the planet in its pristine state.

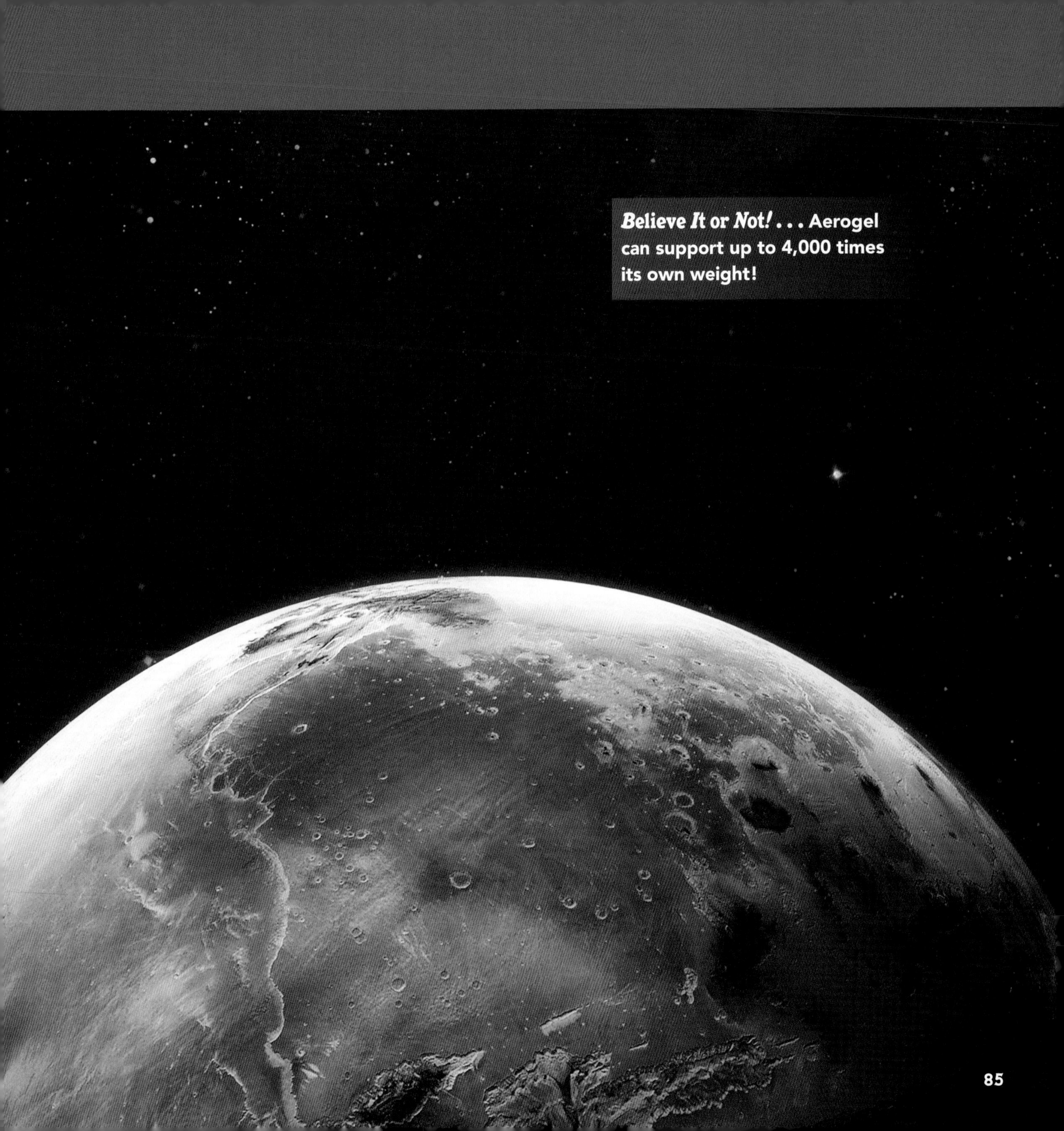

Believe It** or **Not! . . . Aerogel can support up to 4,000 times its own weight!

THE SOUND OF FACES

From the guy who does the voice-over for movie trailers to the announcers on the subway, our lives are full of faceless voices. And while most of us are content to build a mental image of these disembodied orators, a group of MIT researchers has created an artificial intelligence system that can reconstruct people's faces just by listening to their voice.

The application, called Speech2Face, is a deep neural network that was trained to recognize the correlation between voices and facial features by observing millions of YouTube videos with people talking. In doing so, it learned to associate different aspects of the audio waveform with a speaker's age, gender, and ethnicity, as well as certain cranial features such as the shape of the head and the width of the nose.

When the researchers then fed the system audio recordings of people's voices, it was able to generate an image of each speaker's face with reasonable accuracy. However, some improvements are still needed, as the images created by Speech2Face often only bear a general resemblance to the speaker. Nevertheless, faceless voices are one step closer to becoming a thing of the past, which should have major implications for prank callers, at least.

Believe It** or **Not! . . . Two-thirds of the deaths in the Civil War were due to disease!

LET IT GLOW

During the Civil War, the Battle of Shiloh produced more than 23,000 casualties and was the bloodiest battle in American history at the time. Surviving soldiers were then stuck sitting in the dirt and mud for two straight days, waiting for medics to arrive. During a time when the smallest cut or scrape could become infected and kill you, lounging around with battle wounds was not the most sanitary.

When night fell, some of the soldiers noticed their wounds emanating a soft blue glow. Even though a glowing injury would send most of us running for the nearest emergency room today, the soldiers noticed that those with the blue glow had a higher survival rate, prompting this phenomenon to be called "Angel's Glow."

The cause of this heavenly luminescence? Nematodes! A type of roundworm, nematodes are typically a parasitic worm that burrows into larvae bodies and regurgitates bacteria that kills the host from the inside out. This bacteria is called *Photorhabdus luminescens*, and it has a soft blue radiance. The *P. luminescens* was responsible for attacking and killing the harmful germs and bacteria from the mud the soldiers were covered in, which is why they had a higher survival rate!

WARNING: OBJECTS ARE LARGER THAN THEY APPEAR

A stroke can have strange and unusual effects on people's perceptions, from changing their sense of smell to hallucinations. In a particularly curious instance of this perception-twisting effect, a man suffered from a stroke and developed micropsia, a visual disorder in which objects are perceived to be smaller than they actually are, like a miniature model.

As reported in a recent case study published in the journal *Neurocase*, a 64-year-old man was admitted to a hospital in the Netherlands with weakness in his left arm. He had also experienced a temporary loss of vision 11 days prior. Shortly after this incident, the man reported looking in the mirror and perceiving himself as 70 percent of his actual size.

The man also told doctors that he thought his clothes would not fit him anymore because they appeared to be too small. The effect was so severe he often worried whether certain corridors were too small for him to fit through.

The researchers weren't certain why this man acquired such an acute form of micropsia. However, they concluded the man sustained damage to his brain's visual processing hub during his stroke. In an attempt to rectify the damage and maintain some constancy in his perception, the visual system of his brain over-compensated and created this effect.

THE MANY COLORS OF LIBERTY

The Statue of Liberty as we know her is a distinctive and iconic blue-green color, but she wasn't always that way.

Believe it or not, when France gifted Lady Liberty to the United States in 1885, she was a brilliant and shiny shade of copper. Over the course of a few decades, a combination of oxidation and pollution caused her to change color. First, the copper reacted with the oxygen in the air to turn into the pinkish-red mineral cuprite. The cuprite continued to oxidize over time into tenorite, which is black in color. When water and sulfur in the atmosphere started mixing with those copper oxides, the statue began to turn her signature green, helped in part by chloride from the sea spray. Now fully oxidized and chemically stable, the Statue of Liberty has been her iconic blue-green color for more than 100 years.

Believe It or Not! **. . . The Statue of Liberty was used as the United States' first electric lighthouse from 1886 to 1902, when it was decommissioned, as it wasn't bright enough!**

THE RAREST EVENT EVER RECORDED

At a dark matter detector in Italy, scientists were lucky enough to observe the radioactive decay of a xenon-124 atom, a process that takes an unfathomably long time.

The detector, called XENON1T, is located deep within a mountain, where dark matter can theoretically permeate but interfering cosmic rays cannot. If any dark matter collides with even just a single atom making up the 7,716 lb (3,500 kg) of liquid xenon housed inside of it, XENON1T is advanced enough to detect it. And even though it wasn't built for it, apparently it is also sensitive enough to notice when a single atom decays.

According to Ethan Brown, a co-author of the report on the rare observance, "It's the longest, slowest process that has ever been directly observed, and our dark matter detector was sensitive enough to measure it."

Believe It or Not! **. . . The Xenon-124 isotope has a half-life of around 18 sextillion (18,000,000,000,000,000,000,000) years—or about 1 trillion times longer than the age of the universe.**

SWAT
OIL

MAY CONTAIN SMALL PARTS

There have been many selfless sacrifices made over the centuries in the name of science. One of the more recent, but not exactly heroic, examples was when a group of doctors swallowed LEGO pieces in order to study the health effects they could have on children.

In 2018, six doctors with the *Journal of Paediatrics and Child Health* each swallowed a LEGO minifigure head and monitored their digestive health. None reported any problems and passed the heads in an average of 1.7 days.

Taking into account the Found and Retrieved Time (FART) score and Stool Hardness and Transit (SHAT) score, they feel comfortable telling worried parents of toy-hungry children to just "wait and see" when a LEGO is swallowed. They do tell parents, however, to make sure nothing is caught in their child's throat and to seek medical attention if the object is sharp.

FELINE FERTILIZER

In ancient Egypt, mummified cats were considered a reverent gift to the gods. Literally millions of felines were raised for the sole purpose of mummifying them. Later, many of the sacred objects were turned into dirt to fertilize crops.

In 1888, a farmer digging in the sand found an enormous deposit—hundreds of thousands of cats! This was clearly a place for leaving sacrifices. But what should a person do with such a massive collection of cat mummies? A single body may have important information or collector's value, but when there are too many of one item on the market, the item loses value. (This was before Western civilization began to appreciate the cultural and historical value of mummies.)

So, some of the nicest-looking mummies were collected and sold, going to museums and private collections. However, about 180,000 of the cats were sent to Liverpool, ground up, and put into the earth to fertilize crops. The bodies and even wrappings were filled with nutrients that microorganisms could eat and turn into rich soil.

However, it appears this feline-to-fertilizer conversion was a one-time occurrence, as it resulted in a cholera outbreak when the mummies were contaminated in transit to Europe—a mummy's curse, perhaps?

Believe It** or **Not! **. . . In the 19th century, it was common to hold mummy unwrapping parties!**

TURNING TO STONE

Decorated with teddy bears, bicycles, and other souvenirs, Mother Shipton's petrifying well has an unusual quality—it can turn objects to stone!

According to lore, Mother Shipton was born Ursula Southhell in 1488, in a Knaresborough, England, cave. She was said to be a witch and an oracle, associated with many tragic events in the area and predicting, in prose, the horrors that were to doom the Tudor reign. She is to blame for bewitching the well.

The well's petrifying properties can also be explained by modern science. When the well water flows over objects, its unusually high mineral content hardens them—similar to how stalactites and stalagmites form in caves. Astoundingly, objects are hardened in just three to five months!

Believe It or Not! . . . Not all Swiss cheese has holes in it!

CAN'T BE BEAT

According to researchers in Switzerland, not only does music affect the flavor of maturing cheeses, but hip hop produces the best.

Sound bonkers? Well, yes. But, also, not so fast. Sonochemistry is a real field of scientific exploration looking at the influences of sound waves and the effect of sound on solid bodies, using ultrasound to alter chemical reactions. And cheese is essentially a chemical reaction, so perhaps sound waves can have some kind of influence.

To test out this theory, researchers and a cheesemaker placed nine 22-lb (10-kg) Emmental cheese wheels in individual wooden crates in a cheese cellar, and each was played a different 24-hour loop of one song using a mini transducer that directed the sound waves directly into the cheese.

Six months later, the cheeses were analyzed by food technologists, followed by a blind taste test by a panel of culinary experts. The result? Hip-hop cheese topped them all in terms of fruitiness and was the strongest in smell and taste—with the panel also concluding that a clear difference could be discerned between the cheeses. No word on if leaving a speaker on in your fridge will make those processed slices taste any better, though.

SHARP DIET

Pica is the name given to a wide variety of eating disorders that involve the ingestion of items with no nutritive value, such as soil, hair, wood, and metal. The umbrella term covers a range of conditions, including acuphagia, which refers to the eating of sharp objects.

A recent, and extreme, example of acuphagia is Bhola Shankar, who turned up at a hospital in the northwest Indian state of Rajasthan in May 2019 complaining of stomach pains. An initial X-ray revealed a mass of nails lodged in the 43-year-old's stomach, which were then surgically extracted the following day.

There were a total of 116 iron nails removed, each 2.5 in (6.5 cm) long. Shankar was unable to explain why he had eaten them. Usually, doctors first test for a deficiency in iron, which can lead to a compulsive drive to compensate for this by eating metal. However, in many cases, the condition is driven by psychological rather than nutritional causes, and a range of different therapies are needed to treat it.

Believe It or Not! . . . There are estimated to be more trees on Earth than stars in our galaxy!

TWINKLE, TWINKLE, LITTLE NOPE

If you left the Earth's atmosphere and looked at the stars, they wouldn't twinkle. That's because that signature shimmer isn't a property of stars—it's a property of our own atmosphere.

There's even a technical term for it: astronomical scintillation. Our atmosphere is made up of several different layers, each with its own temperatures, densities, and other variables that cause the light from faraway stars to bend and refract. Planets don't appear to twinkle from our point of view on Earth because they are much closer to us and appear larger, making the changes in light caused by our atmosphere not visible to the human eye.

POSITIVELY RADIANT

Believe it or not, radioactive uranium glassware was once all the rage. Collectors today search for these glassware pieces, commonly known as Vaseline glass due to its pale yellow-green color. This uranium glass was made with anywhere from 2 to 25 percent of uranium oxide.

When viewed under a black light, the glassware glows bright green! Don't worry, it doesn't glow based on how radioactive it is; the glass radiates simply because it's made with uranium. The levels of uranium are low enough to make the glass safe enough to handle, but you wouldn't want to eat or drink anything out of it!

ENVIRONMENT

OCEANIC DIAMONDS

Some diamonds have salt water trapped inside them. The origin of that water has been theorized but never tested—until now. A group of researchers in Germany mimicked extreme conditions in a lab, bolstering the theory that the salt water inside some diamonds comes from marine sediment.

The theory holds that when parts of the seafloor rapidly glide under a continental plate (a process called subduction), sediment on the seafloor drops hundreds of miles into the Earth's crust, where high temperatures and intense pressure compound the minerals into small crystals. These then melt in the ancient mantle at temperatures of more than 1,500°F (800°C). These small carbon-fixed stones mix with volcanic magma and spout back onto the Earth's surface as diamonds. Basically, the planet eats the seafloor and then spits out diamonds.

IF YOU'RE HAPPY AND YOU KNOW IT

Tyrannosaurus rex is terrifying in *Jurassic Park*, yes, but what about those sad, stumpy little arms? Well, it looks like Rexy will get the last laugh, as research suggests those arms may have been much more useful (and dangerous) than once thought.

Scientists from Stockton University theorize T. rex's feeble arms were much, much more flexible than we've long thought. T. rex (and other theropods) were probably able to rotate their palms upward and inward, with a range of motion great enough to allow them to perform a clapping motion. Applause might be a concept lost on dinosaurs (with the exception of Barney, of course), but this is still super impressive.

It may not sound like very much, and it's speculative, but there's potential for this to be a huge dino-deal. One possible conclusion to be drawn here is that the ability to rotate the arms and bring them inward toward the chest actually was key in gripping and biting prey.

Believe It or Not! **. . . More time separates the *Stegosaurus* from the *Tyrannosaurus rex* than the T. rex from humans.**

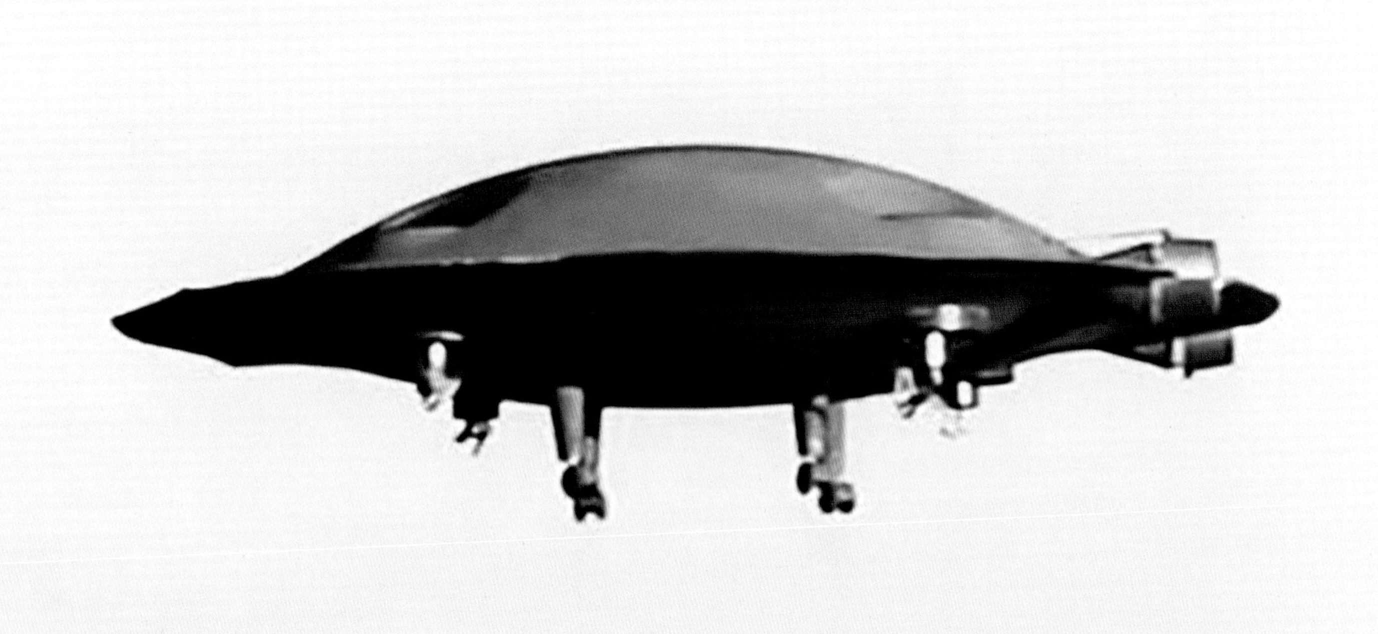

I WANT TO BELIEVE

We finally have the first absolutely 100 percent confirmed sighting of a fully functional flying saucer. A team of scientists and engineers in Romania has developed a first-of-its-kind, hyper-maneuverable flying saucer.

Demonstrating an ADIFO (All-DIrection Flying Object) prototype, the team says a full-scale model would create "a new and revolutionary flight paradigm," and fly like the UFOs depicted in fiction for decades.

At low altitudes and speeds, the saucer uses ducted fans, as you'd see in a normal quadcopter. At higher speeds and altitudes, it uses jet engines loaded at the rear of the saucer. The vehicle also has lateral nozzles, which allow it to perform sudden vertical or horizontal movements at high speeds, much like flying saucers in sci-fi movies. They claim that by using all the methods of thrust simultaneously, ADIFO can perform unique maneuvers "superior to any known aircraft," including sudden stops at speed while maintaining altitude, and full maneuverability while flying upside down.

THE LOST AND FOUND REMAINS OF JOSEPH MERRICK

After nearly 130 years, the remains of Joseph Merrick, also known as "The Elephant Man," have been found—somewhat thanks to infamous Victorian killer Jack the Ripper.

Merrick was born in Leicester in 1862. He started to develop unusual symptoms at five years old, characterized by large abnormal growths across much of his skin and bone, possibly caused by a combination of Proteus syndrome and neurofibromatosis. At the age of 17, he joined a "freak show" that toured across Europe as part of a circus.

Merrick died on April 11, 1890, aged 27, after becoming asphyxiated by the weight of his own head, apparently after trying to lie down. His skeleton has been stored at the Royal London Hospital ever since, but the location of his soft tissue was never officially logged.

While carrying out research for her biography about Merrick, *Joseph: The Life, Times & Places of the Elephant Man*, author Jo Vigor-Mungovin noticed the link between Jack the Ripper and Merrick. She noted that many of Jack the Ripper's 1888 victims were killed in the same district of London where Merrick died two years later. This led her to the records of the City of London Cemetery and Crematorium, where two of the Ripper's victims were laid to rest. Vigor-Mungovin searched the cemetery's records, found Merrick's name, and was able to narrow her research down to a single unmarked grave, where she is "99 percent certain" the Elephant Man's remains were laid to rest.

BABIES... IN SPACE!

Scott Solomon, Professor of Evolutionary Biology at Rice University, has speculated that if human beings ever procreate in space, the resulting offspring might look like your typical Hollywood alien.

In order to spare the child bearer's zero-gravity-weakened bones, space babies would likely be delivered via Caesarean sections. And since the size of our heads is restricted by the size of the birth canal, the increased use of C-sections could lead to our descendants' heads being larger due to the lack of restriction.

Gravity, and the lack thereof, also affects the fluids in our bodies. Normally on Earth, all of the fluids in our bodies are pulled downward. Since this is not possible in space, space babies might develop bloated bodies and puffy faces. Their blood pressure would also increase in the upper body due to zero gravity, causing their eyes to bulge.

On top of all this, space babies might also have a new type of skin pigment. Without Earth's protective ozone layer, radiation from the sun could have disastrous impacts on humans' health, especially our skin. This could lead to a potential change of skin color as evolution tries to counteract the harmful cosmic rays.

All of these factors combined leave us with a baby looking something like the alien emoji.

BLUE BLOODS

Thanks to a natural immuno-response in horseshoe crab blood, biologists have been able to test medicine and health care equipment for contamination since the 1970s.

Having crawled this Earth for more than 400 million years, horseshoe crabs are considered living fossils. They've been around for so long, they actually predate the first dinosaurs by about 200 million years.

Their circulatory system is completely open, meaning their blood doesn't pass through veins or capillaries but circulates around the entire cavity of their body. Horseshoe crabs don't have white blood cells to fight off infection. Instead, their blood is able to detect toxins and bacteria, and then form a robust gel casing around anything harmful.

While most animal blood is red due to a reliance on iron, horseshoe crab blood is rich with copper, making it blue. In the 1970s, scientists developed a way to use their blood to validate the purity of medicine and equipment. Prior to this, infections caused by attempts at medical care were sometimes worse than the injury itself. Nowadays, there is a synthetic substitute for horseshoe crab blood, which may soon make horseshoe crab milking obsolete.

Believe It** or **Not! **. . . The rarest blood type is RH-Null, or "golden blood." Less than 50 people in the world have it!**

BOLD MOLD

Slime mold, despite its name, is not actually a mold, a fungus, a plant, an animal, or a bacterium. It belongs to a kingdom of life called Protista that contains any single-celled organism that is not an animal, plant, or fungus.

One species of slime mold is called *Physarum polycephalum*, which means "many-headed slime." It's not a single creature but rather a collection of unicellular organisms that can band together in a single form. They can be chopped into many pieces, only to fuse back together within a few hours. If they come across any other slime molds along their travels, they will join together.

There is neither a male nor a female of the species, but more than 720 different sexes. It can creep from place to place at up to 1.6 in (4 cm) per hour by extending stringy fingerlike protrusions.

To get even weirder, slime mold has neither a brain nor a central nervous system, nor any neurons. Nevertheless, some scientists argue they exhibit intelligence because they can "learn" from experience and change their behavior accordingly.

If artificial intelligence doesn't overtake mankind, slime mold just might.

THE MIGHT OF LIGHT

Light might have no mass, but it can still push things around. This is known as radiation pressure. Light particles (photons) carry a momentum with them, but how this momentum is transferred is not exactly clear. However, new research has come up with a way to actually study these interactions between light and matter.

An international team constructed a very special experiment to study the momentum of light. Photons carry a tiny momentum, and their effect can only be studied cumulatively. Still, there were no devices sensitive enough to measure the effect. This is why it has been so difficult to study how radiation pressure is converted into force or movement.

The team built a mirror fitted with acoustic sensors. They shot laser pulses at the mirror and studied the effects. Then the sensors recorded the vibration generated by the photons. The elastic waves moved across the mirror like ripples on the surface of a pond. These observations finally confirm predictions regarding the transfer of momentum.

WHEN EVERY DAY IS UNFORGETTABLE

Think of a random date from your childhood onward. Can you remember what day of the week it was, what you were doing that day, and what news events occurred? Somebody with an ordinary memory would probably have trouble, especially if nothing significant transpired. But 34-year-old Joey DeGrandis of New York can almost definitely tell you within seconds.

DeGrandis is one of fewer than 100 people worldwide with Highly Superior Autobiographical Memory, or HSAM, a rare condition that typically allows them to recall life memories from specific dates with great accuracy and ease.

Take February 25, 2010. Your average person probably can't tell you about what they did on that exact day. But DeGrandis remembers that it was a Thursday (correct) and it was snowing in New York (also correct). He knows he went to Katz's Deli with his friends Jeff and Dan, and it was two days after he had gone on a job interview for a position he really wanted.

HSAM manifests itself differently in those who have it. Some view it as merely an interesting ability. Others say it has caused significant challenges in their daily lives. Some experts think HSAM could open new doors in the study of memory and potentially offer new insights into Alzheimer's disease and other memory loss issues.

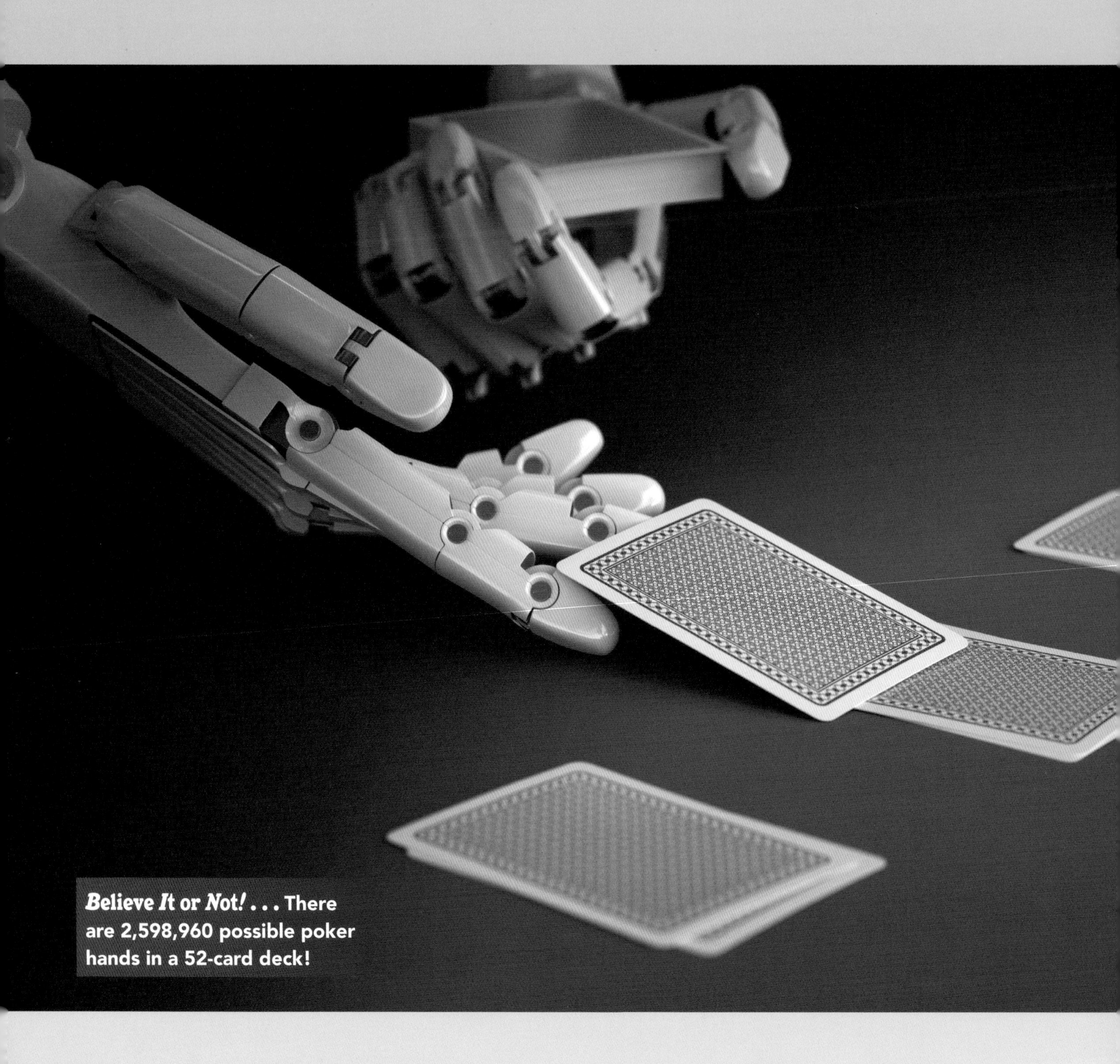

Believe It or Not! **. . . There are 2,598,960 possible poker hands in a 52-card deck!**

UPPING THE ANTE

Artificial intelligence can beat the world's best at chess, Go, Jeopardy!, and now six-player no-limit Texas hold 'em poker—showing it's well on track to become our almighty overlord in the not-so-distant future.

Programmers have developed a computerized poker champion that has successfully defeated Darren Elias (who holds the most World Poker Tour titles), Chris "Jesus" Ferguson (winner of six World Series of Poker events), and more than a dozen pros who, between them, have won over $1 million from the game.

One thing that makes this triumph so special is the secretive nature of poker. In chess and Go, both players can see everything that goes on the board. In poker, they can't—cards in play aren't always visible and players can bluff. This, the researchers say, makes it a trickier game to play for a machine built on logic and probabilities.

The machine, "Pluribus," uses a limited-lookahead search algorithm, enabling it to predict the strategy its opponents will use in the next two to three plays (as opposed to the entire game). Pluribus also thrives on unpredictability. After all, it wouldn't get very far if it saved its bets for excellent hands only. The novel strategy makes AI more relevant to "real-world" problems, which often involve missing information and multiple players.

DON'T STRESS

Space travel may be out of this world, but the side effects can be unexpected and dangerous. Certain dormant viruses have been seen resurrecting inside astronauts' bodies after time spent in space. Specifically? The herpes virus.

NASA scientists have discovered that the herpes virus reactivates during longer stints of space travel. The virus has co-evolved with humanity and only wakes up during times of high stress or illness. Immuno-suppressing stress hormones, like cortisol and adrenaline, surge during spaceflight, simulating a high-stress environment for the dormant virus to resurface.

Four of the eight herpes viruses known to infect humans have been identified in astronauts, such as the ones responsible for chicken pox and mono. Once infected, the astronauts stay contagious for up to 30 days after returning from space!

INVINCIBLE INSECTS

It is often said that cockroaches could survive a nuclear apocalypse—they can't. But that doesn't mean they are easy to get rid of. And, lucky for us, these creepy crawlies are only getting harder and harder to kill.

Researchers recently tested out different treatments of three insecticides in numerous cockroach-infested apartments across Indiana and Illinois over six months. Regardless of the different chemical cocktails, the researchers were unable to reduce the size of the cockroach populations. (Some even got larger.) Even scarier, some populations developed resistance to the insecticides within a single generation.

While cockroaches are still not immune to a good foot stomping, this new study does suggest that humans need to wise up when it comes to pest control. The researchers say their findings highlight the need for combining chemical treatments with traps, improved sanitation, and vacuums to control cockroaches, rather than just relying on insecticides.

DEADLY DIET

It is said that you should eat everything in moderation, but "everything" probably shouldn't include microplastics. The average American consumes 1,314,000 calories, almost 80 lb (36 kg) of sugar, and more than 74,000 microplastic particles every year. These particles are micro-sized pieces of plastic from larger pieces that have been broken down.

Researchers have reviewed 26 studies analyzing the number of microplastics found in fish, shellfish, added sugars, salts, alcohol, tap or bottled water, and air to find out how much microplastic we really consume. Varying factors, such as age, sex, and dietary preferences affect the overall number, but the general estimate? Anywhere between 74,000 and 121,000 particles a year. If you like to drink bottled water, add an extra 90,000! The food we eat is responsible for 39,000 to 52,000 particles, while we inhale the rest throughout the day.

Though these numbers may sound alarming and negative side effects are possible, scientists say there is no need to worry about imminent danger.

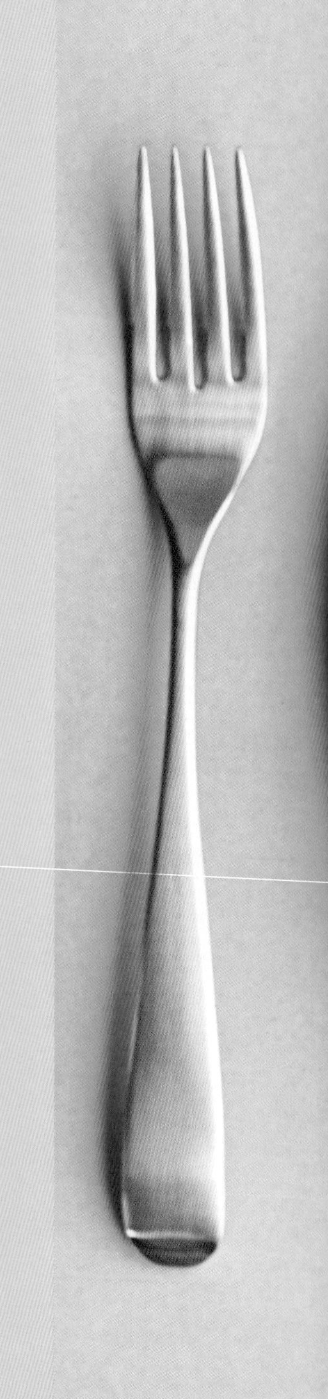

A GLOWING LEGACY

Famous physicist and chemist Marie Curie made significant contributions to science, including coining the term "radioactivity," but if you want to handle her personal manuscripts, you'll have to wear protective clothing and sign a liability waiver.

The Polish-French scientist worked extensively with radioactive substances before the dangers of radiation were understood. She even walked around with bottles of polonium and radium—elements she and her husband, Pierre, discovered—in her pockets and kept bottles of them on her shelves, admiring the dim glow they gave off in the night.

Curie passed away in Savoy, France, on July 4, 1934, from aplastic anemia, thought to have been caused by her excessive exposure to radiation. Her papers will remain radioactive for more than 1,000 years and are currently housed in lead-lined boxes at the Bibliothèque Nationale in France.

Believe It or Not! **. . . Marie Curie was the first person to receive two Nobel Prizes—first in physics, then later in chemistry.**

BEANS ARE GO

SPACE SUPERSTITIONS

You wouldn't think that the world's foremost scientific agency would be very superstitious, but NASA has some long-standing traditions that they live and breathe by.

Before every launch from the Kennedy Space Center, astronauts eat a meal of steak, eggs, and cake, no matter the hour, and must play poker until the commander loses. They typically also bring along a stuffed animal. Not only is the plush considered lucky, but its fluffy body is a safe enough object to signal when gravity shifts and objects start floating.

And it's not just the astronauts. Launchpad engineers believe it's bad luck for astronauts to see the rocket before launch day and take great measures to ensure none of them see their ride during the sometimes days-long journey to the launch pad. And back at mission control, every shuttle launch is celebrated with a round of baked beans.

Cosmonauts have their rituals as well. Every single person launched into space from Baikonur Cosmodrome in Kazakhstan has peed on a bus tire just before takeoff. This unpeelievable tradition started after Yuri Gagarin—the first man to orbit the Earth—took an emergency pit stop on a bus tire in 1961.

THE POISON GARDEN
THESE PLANTS CAN KILL
THESE PLANTS CAN KILL

FOR YOUR EYES ONLY

Alnwick Castle in Northumberland, England, plays host to the small but deadly Poison Garden—filled exclusively with around 100 toxic, intoxicating, and narcotic (illegal) plants.

Surrounded by the beautiful 12-acre Alnwick Garden, the boundaries of the Poison Garden are kept behind intricate black iron gates. The only way to see the garden is on a guided tour, during which visitors are strictly prohibited from smelling, touching, or tasting any plants. (Although some people still occasionally faint from inhaling toxic fumes while walking in the garden.)

The garden contains exotic plants like the *Brugmansia* of Brazil, which can cause paralysis and death, along with common English plants like laurels, whose smell can render people unconscious. Other toxic plants featured include hemlock, castor beans, and deadly nightshade. The garden also contains a variety of illicit drugs, including cannabis, coca, and poppy.

The garden is the brainchild of Jane Percy, Duchess of Northumberland, who was tasked with finding something to do with unused castle grounds. Despite its royal origins, this is one garden where you *don't* want to stop and smell the flowers.

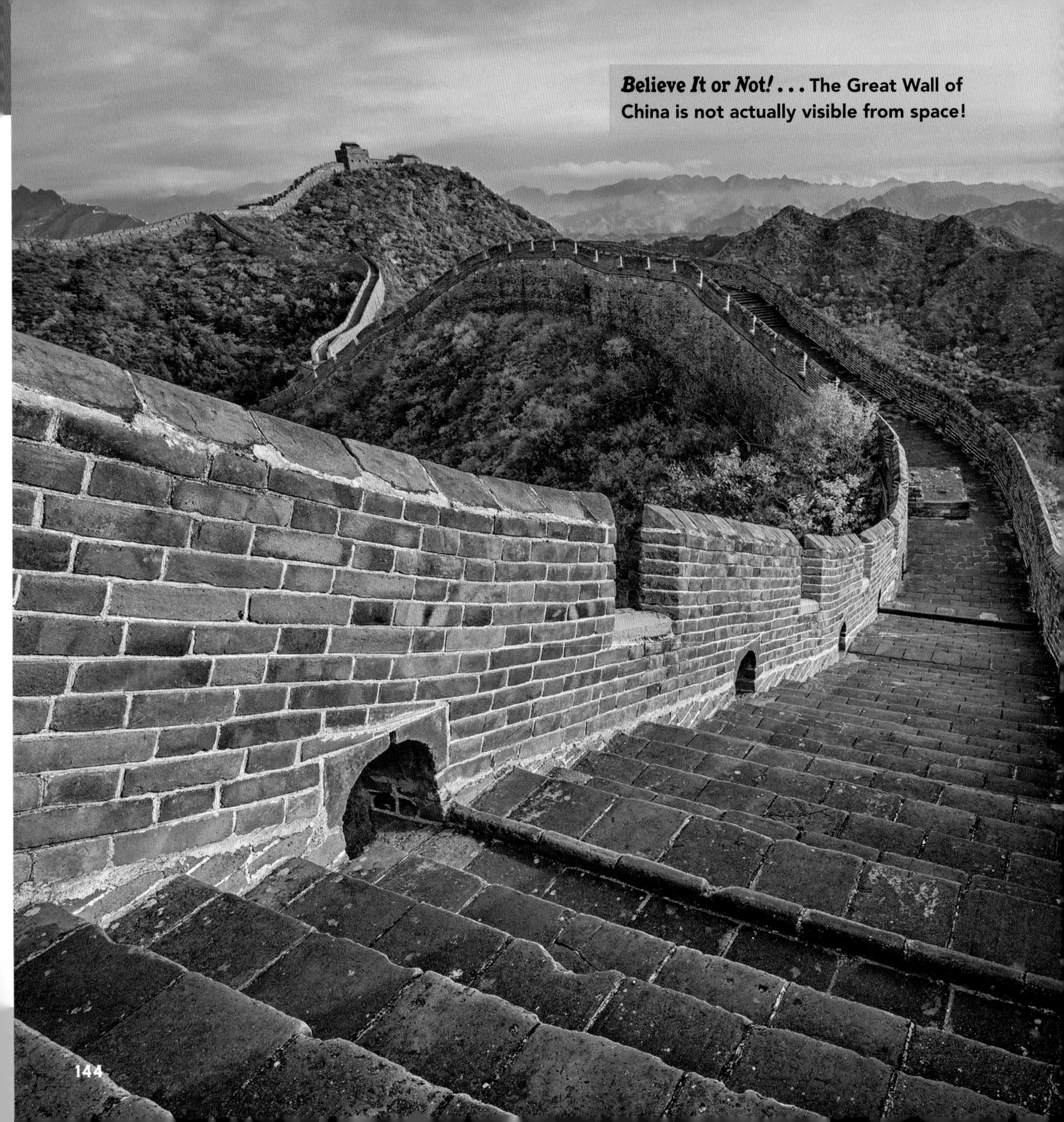

Believe It or Not! **. . . The Great Wall of China is not actually visible from space!**

BRICK BY BRICK BY GRAIN

Centuries-old portions of the Great Wall of China are still standing thanks to a powerful mortar created with an unexpected ingredient—sticky rice!

In one of the greatest technological innovations of the Ming Dynasty (1368–1644), workers developed sticky rice mortar. They crafted it from a mixture of slaked lime—a standard ingredient in mortar—and sweet rice flour. The result was the first composite mortar in history, a potent mix of inorganic and organic ingredients.

A team of researchers from Zhejiang University investigated the chemical composition of Ming-era mortar and found that the legendary strength of rice-lime mortar comes from amylopectin. When amylopectin (the organic portion of the mortar recipe) comes into contact with calcium carbonate (the inorganic part), a complex interaction occurs. Acting as an inhibitor, the amylopectin controls the growth of the calcium carbonate crystal.

The result is a more tightly bonded mixture with three key advantages: First, it's highly water resistant. Second, it shrinks less and holds its shape. Third, the key chemical reaction in the mortar continues over time. Put another way, the mortar gets stronger as the years go by!

R.I.P.

WHERE SATELLITES GO TO DIE

Have you ever wondered what happens when a spacecraft launched into space shuts down and begins falling toward Earth? They don't just drop randomly; otherwise, the thousands of satellites, landing modules, experimental spacecraft, and space stations would be coating us with debris every week.

Located 2,500 mi (4,023 km) off the coast of New Zealand, a spacecraft graveyard lies in what's officially called the South Pacific Ocean Uninhabited Area (SPOUA).

SPOUA has been designated as entirely void of human life; it contains no islands and very few shipping lanes. Scientists determined it is the least likely place for an incoming spacecraft to harm a human being. At the center of SPOUA lies Point Nemo, the exact place on Earth that is farthest from any land mass. Out of sight, out of mind, right?

NO MORE ICE, ICE, BABY

Nothing makes mornings worse than they already are quite like having to scrape a night's worth of ice off of your windshield.

So, if you don't want to waste energy using a scraper or waste time with that defroster heating up your windshield, then science has got a quick and cheap solution for you!

The solution is an actual chemistry solution—a mixture of one part water and two parts rubbing alcohol that will dissolve the frost from the windshield rapidly and effortlessly. Isopropyl (rubbing) alcohol has a freezing point of –128°F (–89°C), so the solution not only doesn't freeze, it actually helps lower the freezing point of water and melts the frost away.

Looks like whoever owns this car is going to need a lot.

SUDDENLY SAGE

Sudden savant syndrome, also known as an acquired savant, is the inexplicable onset of astonishing new abilities at a gifted or prodigal level. There is no scientific explanation for why this occurs.

One such case of a sudden savant is a 28-year-old man from Israel. His only musical experience was memorizing a major and minor chord, without fully understanding how musical scales worked. While playing random keys at a piano in a mall, he suddenly began playing popular songs just from his memory. He knew every chord and scale, how the notes worked together, and how to recognize harmonies.

With no history of a neurological disorder or brain injury, this man became a sudden musical savant. Other cases of sudden or acquired savant syndrome have seen people become prolific painters, photographers, and even mathematical geniuses. Some of these instances were preceded by head injuries, but we don't recommend trying to recreate them in hopes of becoming the next Mozart. Just practice like the rest of us.

URINE FOR A SURPRISE

Despite repeated assertions in popular culture and across online forums throughout the years, your pee isn't actually sterile.

This "fact" sprang from the development of urinary tract infection tests by epidemiologist Edward Kass. He began testing surgery patients' urine and cleared them if they got a "negative" for bacteria. This led many people—even doctors—to believe that healthy urine is completely free from bacteria. This, however, was an oversimplification of the test; a negative result doesn't mean there aren't any bacteria at all, but instead that the bacteria falls below a certain threshold. Everything in or on the human body is contaminated with microbes.

But don't worry—just because your pee contains microbes doesn't make it dangerous. Much of the bacteria in the human body is essential to our bodily functions, especially digestion. So despite containing germs, your pee is not dangerous unless you have something like a urinary tract infection, but we still wouldn't recommend drinking it.

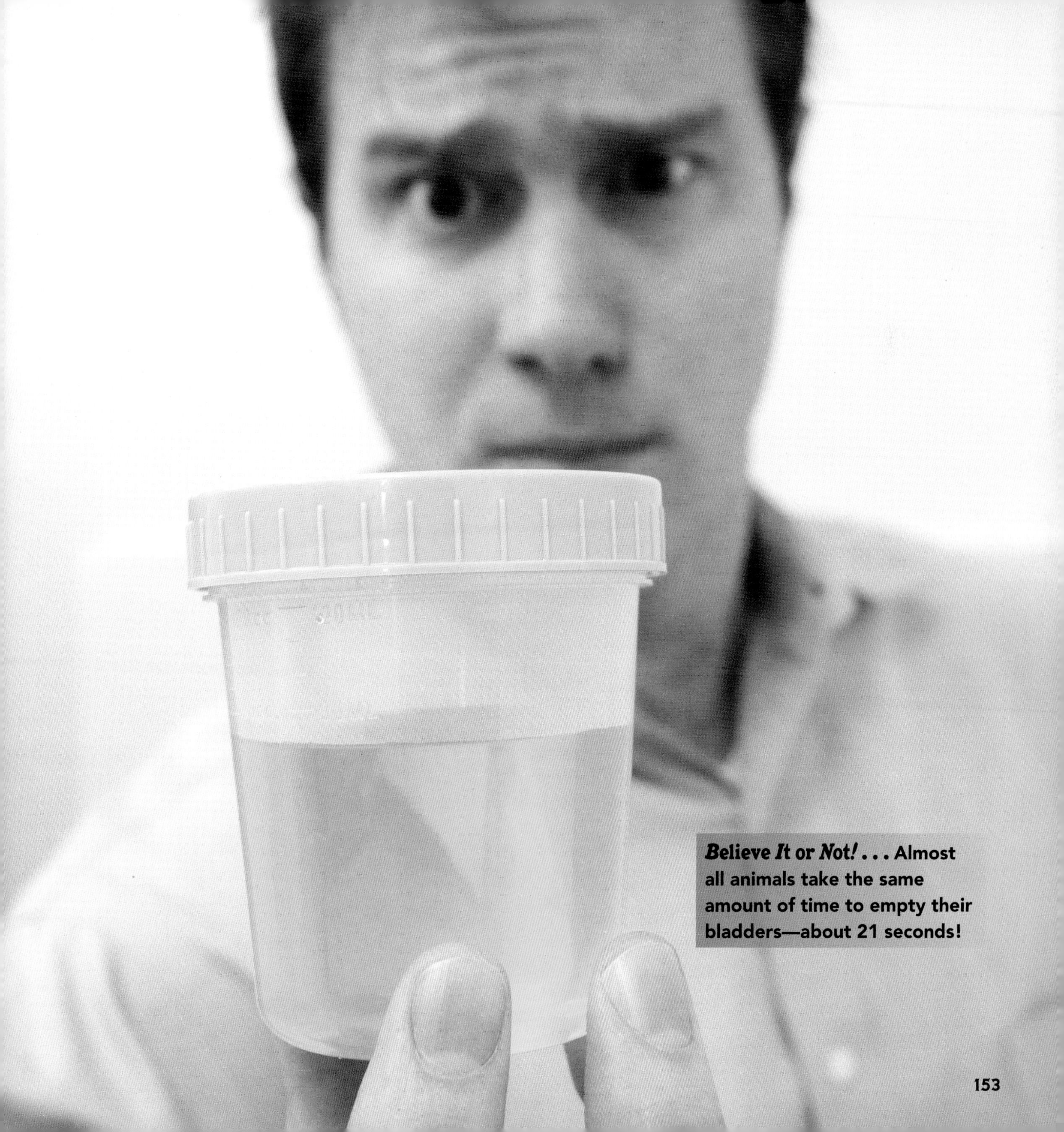

***Believe It* or *Not!* . . .** **Almost all animals take the same amount of time to empty their bladders—about 21 seconds!**

mazda
JND-43-76

OH, HAIL NO

A freak hailstorm in Guadalajara, Mexico, on June 30, 2019, left residents of at least six neighborhoods to find the streets, and their vehicles, buried in up to 5 ft (1.5 m) of ice. It was all the more shocking because up until then, the city had been basking in temperatures of 88°F (31°C).

Seasonal hailstorms have been known to occur, but nothing on this scale has been seen in recent memory. While people seemed to enjoy the novelty of a snow day in Guadalajara, not everyone was pleased. At least 200 homes and businesses reported hail damage, and it's thought at least 50 vehicles were swept away by ice running down from the hills or were buried under hailstones, too.

QUANTUM CANVAS

Humans love art. From frescos in the Sistine Chapel to doodles on notepads, we are constantly letting out our artistic side. And now art can enter a new frontier: the quantum world. Researchers have developed a way to "paint" tiny masterpieces on a special state of matter called a Bose-Einstein Condensate (BEC).

The BEC happens when a diluted gas of certain particles—in this case, rubidium atoms—is cooled down to a few billionths of a degree above absolute zero. Under these conditions, where most atoms are in their lowest possible state, microscopic quantum phenomena suddenly become macroscopic.

And on this quantum blob of matter, researchers were able to project an image using a laser to create a "light stamp." They were able to create micro-copies of paintings such as the *Mona Lisa* and *The Starry Night* by Van Gogh. These pieces of art are tiny but not atom-sized. They are roughly 100 microns across, which is more or less the width of a human hair.

WARNING: ZOMBIES AHEAD

We may not have to worry about a zombie virus infecting humans, but gastropods, such as snails, should keep their antennae up.

The *Leucochloridium* is a parasitic flatworm ingested by snails in the form of eggs found in bird droppings. Once consumed, the eggs hatch into larvae and "infect" the snail by forcing it to move into an unsafe environment, such as a sunny area or higher ground. The worm moves into the snail's transparent eyestalks and pulsates, mimicking a caterpillar, to attract hungry birds.

The snail may not want to be eaten, but the worm has other plans. Once inside the bird's gastrointestinal tract, the zombie snail infects it with the *Leucochloridium* larvae that then mature and release eggs. After the bird excretes these eggs, the cycle continues once again.

***Believe It or Not!* . . .** The victim snail will often survive this horror show, as a hungry bird usually just plucks off the offending eyestalks, leaving the snail to regenerate the lost body parts and potentially become part of a parastic flatworm's life-cycle once again.

BIG BANG BUZZ

Some of the static on analog TVs and FM radio stations is from the Big Bang!

According to the theory, the universe began its life around 13.8 billion years ago, when literally everything erupted from a singularity, causing a massive release of energy. You might think that energy would disappear over time, but since the universe contains literally everything that exists, it just bounces around the cosmos.

Now it can be observed as background radiation called the Cosmic Microwave Background (CMB). About 1 percent of the static between channels on analog TVs and the white noise on empty radio stations is caused by the afterglow of the Big Bang. The switch to digital TVs and streaming music has made experiencing the CMB in these ways a much rarer occurrence.

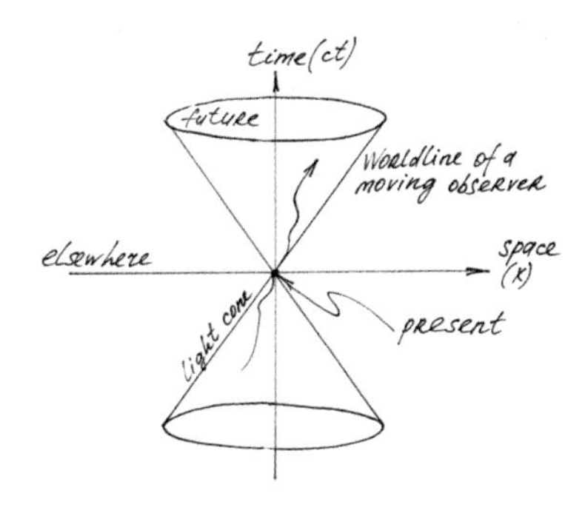

TECHNOLOGY

A DIFFERENT KIND OF CATCALLING

In 1929, two Princeton University researchers were able to turn a living cat into a working telephone.

In an effort to learn more about how the auditory nerve perceives sound, Professor Ernest Wever and his research assistant Charles Bray removed part of a sedated cat's skull and attached one end of a telephone wire to the feline's right auditory nerve and another to a telephone receiver. With Wever 50 ft (15 m) away in a soundproof room with the receiver, Bray talked into the cat's ear and, astonishingly enough, Wever was able to hear him loud and clear!

Although some aspects of the experiment were later disproven, it is believed their work inspired research that helped develop the cochlear implant—a device that stimulates the auditory nerves of deaf individuals by converting sound into electrical signals.

Believe It** or **Not! **. . . The World Chess Boxing Organization motto is "Fighting's done in the ring, and war's waged on the board."**

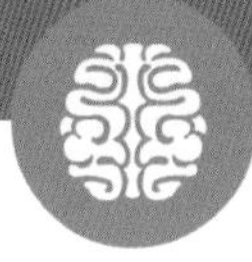

BRAINS AND BRAWN

There is no finer sport that matches wits and fists than chess boxing. Whether you choose to call it a hybrid fighting sport or hybrid board game, chess boxing is exactly what you think it would be. Two opponents play a game of chess for one round, box for another, and then repeat.

The match lasts 11 rounds and begins with chess. The combatants square off in a timed game of chess for six minutes, then switch right over to fighting. Once the boxing round ends, they are back at the board continuing their ongoing chess match. Separate referees typically monitor the chess match and boxing.

Athletes only have a short 60 seconds while the board is being removed from the ring to take a break. A win is decided by a checkmate or knockout. If the chess match ends in a tie, the match is decided by technical points, and if that's a tie, the black chess player wins.

SCALY BANDAGES

In 2016, doctors in Brazil successfully treated severe burns by covering them with dressings made of fish skin.

The skin comes from the abundant and disease-resistant tilapia fish. The alternative bandages go through a process that removes the scales, muscle tissue, possible toxins, and fishy smell before it is stretched, laminated, and cut into strips. These strips can then be placed on damaged skin, such as burn victim Maria Ines Candido da Silva, who was the first patient to receive this treatment. The flexible and moist nature of the fish skin makes it the perfect material for dressing burns and has been increasing in usage.

WHY CAN'T WE WALK THROUGH WALLS?

The question looks like an easy one and the answer fairly straightforward. You are solid. Walls are solid. Therefore, you can't walk through them.

But things get a little more complicated when you look at it on a microscopic level. We—and everything else in the universe—are really just an assemblage of atoms, and atoms are almost entirely empty space. To be pedantic, they are 99.99999 percent empty space. So if this is the case, why can't we walk through walls?

It comes down to the principle that no two fermions (a group of subatomic particles that includes electrons, protons, and neutrons) can be in the same state or same configuration at any one time. Electrons are constantly swarming haphazardly around an atom's nucleus at such a speed that it is as if they are taking up all of the empty space at once. Also, the electrons of one atom repel the electrons of other atoms, so that each atom remains distinct.

This means that if you were to walk through a wall, two electrons (yours and the wall's) would have to coexist in the same space—something that is just not possible. Therefore, despite the fact that we are almost entirely empty space (a mind-boggling fact in itself), we cannot walk through walls—or any other solid material, for that matter.

KELLY
HAM
NYBERG
FOSSUM
HOSHIDE
CHAMITOFF
NOWAK
FOSSUM
REITER
SELLERS
UF-1
ОНУФРИЕНКО WALZ BURSCH
CULBERTSON ДЕЖУРОВ ТЮРИН

SPACED OUT

We have been sending people into outer space since 1961. But what do we really know about what happens to the human body during space travel? Researchers followed the lives of astronaut Scott Kelly and his twin brother Mark, a retired astronaut, to examine what happens to your genetic makeup in space.

During Scott's 340 days on the International Space Station, 10 studies were conducted on both twins. The results? Scott Kelly's genetic expression, bone density, immune system responses, and telomere dynamics were altered.

One of the more intriguing differences between the twins was a shift in Scott's gut bacteria, with an increased production of good bacteria. These levels returned to normal when he was back on Earth. Scott also had longer telomeres—"caps" on the end of chromosomes related to aging—while he was in space, even though they typically shorten as a person ages. After about six months back on Earth, these also returned to normal.

Human DNA chemical compounds remained unchanged during the mission, which makes public space travel a growing possibility!

Astronaut twins Mark (left) and Scott Kelly (right).

FALLING FOR YOU

While the death of a whale can be a sad sign of polluted waters, food scarcity, or human violence, the gargantuan bodies left behind by these majestic creatures can create a hotbed for life.

Sometimes called a whale fall, carcasses that sink to the deep depths of the ocean's abyss are an important transmission vector for nutrients to an otherwise static part of the ocean.

Over the course of months, the soft tissue is completely eaten away by fish, sharks, shrimp, eels, and crabs. The detritus and material of the whale don't just provide direct sustenance, however. The nutrients that make it into the soil give rise to plant life and bacterial mats, which in turn are a new food source for the whole biological community.

Even the remaining skeleton can become a basis for colonization, with microorganisms living off the chemical reaction of its decomposition. Anemones specific to whale falls, as well as unique bone-eating zombie worms, call the rotting carcass home. These organisms can be supported for decades from a single whale carcass.

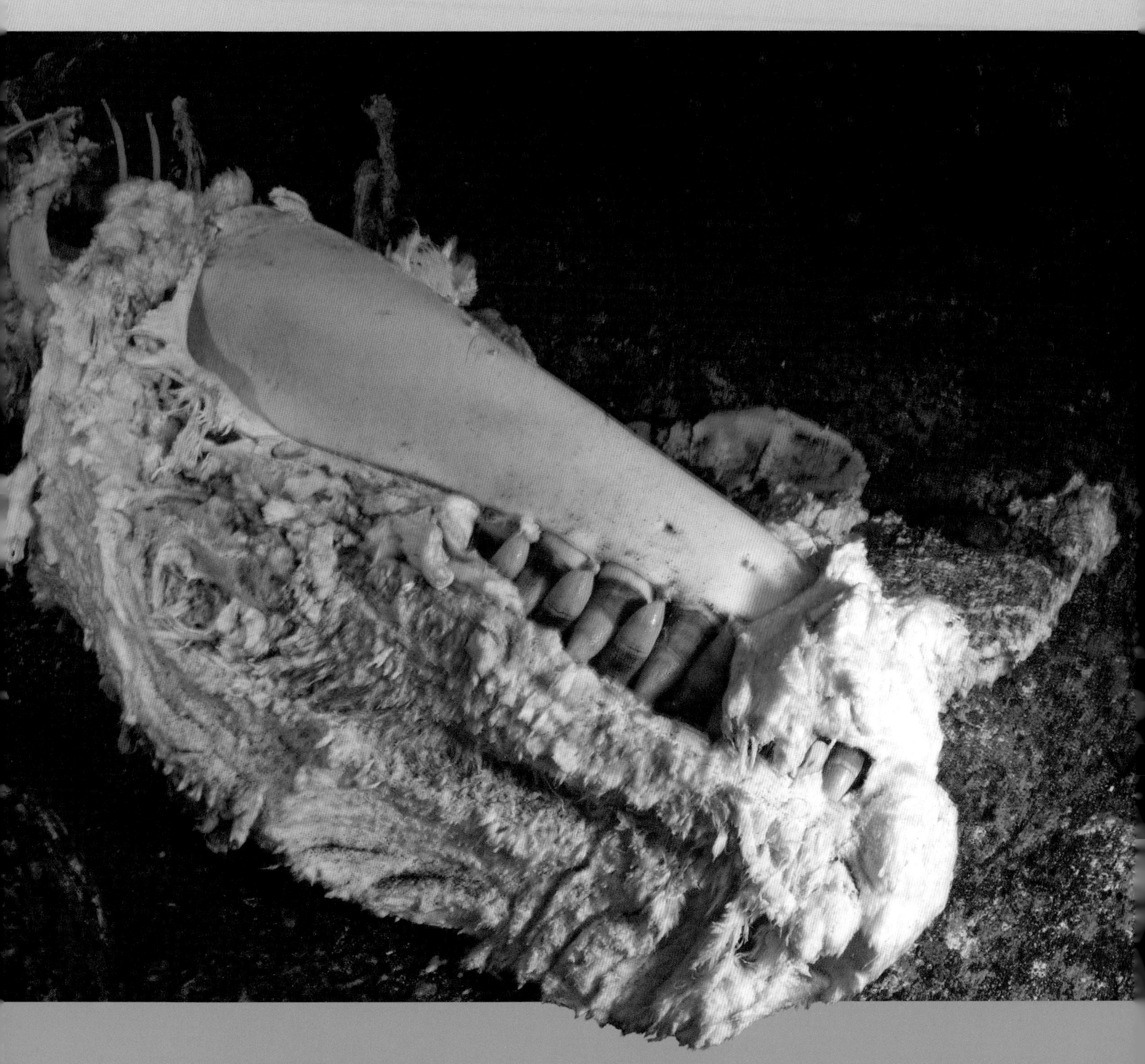

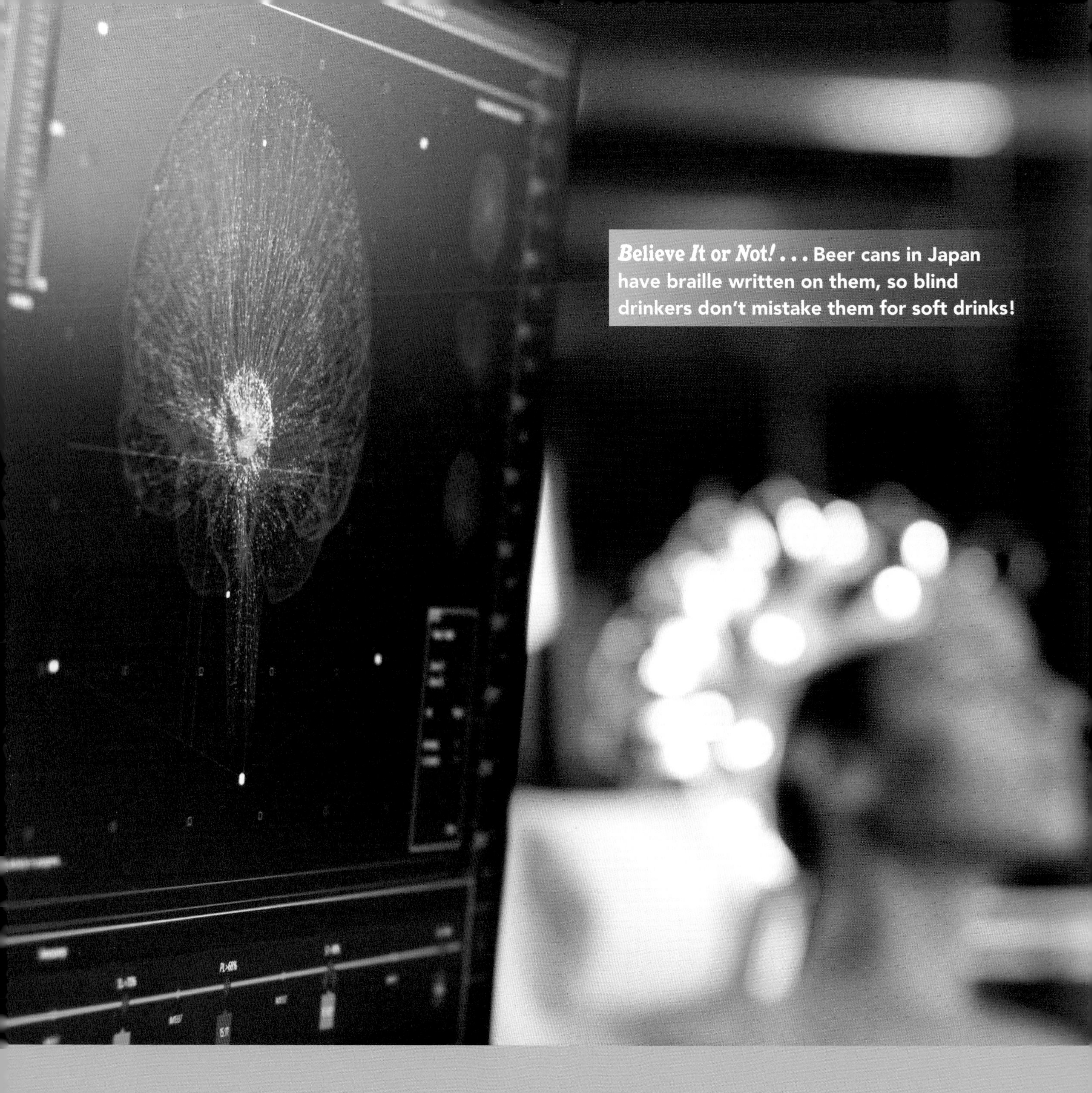

***Believe It* or *Not!* . . .** Beer cans in Japan have braille written on them, so blind drinkers don't mistake them for soft drinks!

DO YOU SEE THAT SOUND?

It's often said that as you go blind, your other senses will be heightened. There has been growing evidence that in blind people, the part of the brain that processes vision repurposes itself to process sound!

A study was conducted that monitored the brain activity of both blind and sighted people as they listened to three different audio recordings. One was perfectly clear, the second was slightly distorted, and the third was almost unrecognizable. While the monitors showed activity in the auditory sections of the brain in both groups, blind subjects showed increased activity in the visual sectors. Researchers argue that the visual parts of the brain are, in fact, being used to process sound!

CREATURES OF THE DEEP

The ocean is a deep, dark, and mysterious place. It's so mysterious that scientists aren't capable of explaining every phenomenon they come across, especially abyssal gigantism!

Stories of mythical sea serpents, such as Jörmungandr of Norse mythology or the Loch Ness monster, have been told for centuries. These supersized aquatic creatures aren't too difficult to fathom when you take a look at abyssal gigantism, when deep-sea creatures grow much larger than their shallow-water family members.

Isopods, a type of crustacean, have been found measuring from just a few micrometers to the size of a small dog! Japanese spider crabs, another type of crustacean, can have a leg span of up to 15 ft (4.6 m)!

Oarfish, the animal that most likely inspired tales of the Loch Ness monster, is the longest bony fish in the sea. It can measure up to 56 ft (17 m) in length!

And fear of the kraken, an enormous fabled cephalopod, is likely based on the giant squid, which wasn't caught on camera until 2004. Believe it or not, they can grow up to 40 ft (12 m)—that's the size of a school bus!

789879
123213
789879
789879

ARE YOU SMARTER THAN A COMPUTER?

Computers certainly aren't dumb. However, for all their brainpower, machines are still surprisingly clunky and awkward in the art of conversation, especially when it comes to answering questions.

To overcome this weakness, computer engineers at the University of Maryland generated a collection of more than 1,200 questions that completely bewilder even the best computer answering systems today, despite being comparatively easy for people to answer, hoping they will become better trained at communicating with humans using language. Here are a select few examples of the trivia questions:

- Name this European nation that was divided into Eastern and Western regions after World War II.
- Identify this metal that is used for decorative coatings and many musical instruments.
- Name this South African leader who became president in 1994 after spending 27 years in prison.

But how can a computer be foiled by such simple questions? The reason has much more to do with language than knowledge. It's notable that the questions are worded in a slightly odd way. That's because they're laced with six different language phenomena that consistently stump computers but don't tend to phase humans.

OH, SNAP!

Did you know that a single strand of spaghetti is referred to as a spaghetto? If you try to snap a spaghetto in half, you'll end up with three unequally sized pieces.

Why can't they break into two equal halves? Once snapped, the strands send a violent ripple backward and cause additional breakage. How can you remedy this aggressive shockwave? You twist it!

Scientists from MIT, Cornell University, and the University of Aix in Marseille have created a device that rotates the spaghetto a full 360° before pushing clamps together to cause a break. The reverberations then release through the unwinding of the pasta, preventing further shattering!

TECHNICALLY THE TALLEST

It's common knowledge that Mount Everest is the tallest mountain on Earth, but what does "tallest" even mean? Height above sea level? Greatest distance from base to peak? What about distance from space? The answers to all of these questions change which mountain earns the tallest mountain title.

Mount Everest stands a proud 29,029 ft (8,848 m) above sea level. But that status delicately hangs on the way sea levels work. Sea levels vary all over the globe, and the elevation of Mount Everest is measured from the world average.

From its base to its peak, Mauna Kea measures more than 33,000 ft (10,000 m). Unlike Everest, the base of this Hawaiian mountain starts way below the Pacific Ocean. If its base were raised to sea level—where the base of Everest is measured—Mauna Kea would stand about a mile taller.

At 20,548 ft (6,263 m), Mount Chimborazo is the tallest mountain in Ecuador. It is also very close to the equator. As the Earth spins, it pushes out around the equator, becoming somewhat disc-shaped. Because of this planetary centrifugal force, Chimborazo is about 1.3 mi (2.1 km) closer to space than Everest.

Believe It or Not! **. . . It is estimated that climbers leave a combined 26,500 lb (12,020 kg) of human waste on Mount Everest every year.**

THE WRIGHT WAY

Brothers Orville and Wilbur Wright achieved the first sustained flight with a heavier-than-air aircraft on December 17, 1903. Orville flew 120 ft (36.5 m) over a period of 12 seconds.

The Wright brothers paved the way for revolutionizing air travel, and on July 20, 1969—just 65 years, 7 months, and 3 days later—the *Apollo 11* spacecraft was sent to the Moon. In this short amount of time, air travel went from a single engine and propeller mechanism to the advanced technology of the *Saturn V* rocket.

The *Saturn V* rocket stood 363 ft (111 m) tall—taller than the Statue of Liberty—and had an on-board computer that guided it into Earth's orbit. NASA had large computers back on the ground to help with navigational corrections.

In between the creation of the first aircraft and the engineering of technologically advanced spacecrafts, airplane passengers in the 1930s sat in wicker seats! We just celebrated the 50th anniversary of the first Moon landing—who knows where we'll be flying off to next.

Believe It or Not! **. . . The Wright brothers' historic flight was initially reported in just four newspapers!**

CHEMISTRY

HOLD IT IN

Hold your cheeks together the next time you go under the knife, because a squeaky-bottomed Japanese woman recently ended up with burns to her legs and waist after her fart caught fire during surgery.

According to a Japanese newspaper, the lady was being treated at a hospital in Tokyo, where doctors were using a laser to perform a procedure on her cervix. A report into the incident explained that the laser ignited gases that leaked from her intestines, causing the surgical drape to ignite.

Farts contain a number of different gases that are released as by-products of our intestinal activities. Among these are hydrogen and methane, both of which are flammable and give our windy expulsions their combustible characteristics.

Believe It** or **Not! **. . . A sperm whale can detect a 1-ft-long (0.3-m) squid at a range of 1 mi (1.6 km) using sonar sound waves.**

CAN YOU HEAR ME NOW?

Brought to fame by Herman Melville's whale-hunting epic *Moby Dick*, sperm whales are some of Earth's most mysterious creatures. Even with the ability to dive thousands of feet deep, researchers still don't know what goes on below the water's surface.

What we do know about sperm whales is that they communicate by using a series of four different clicking noises. These clicks are used for both social communication and honing in on prey. When hunting, the sperm whale's noise reaches up to 230 decibels (dB)!

To put that in perspective, a normal conversation between two people is held at about 60 dB, a lawn mower is at about 105 dB, and a sports arena ranges from 120 to 130 dB. Any sound higher than 85 dB is considered harmful, but your eardrum can rupture at 150 dB!

OXYGEN
GENERATION
EFFICIENCY
74%
G7

THE FERTILE FRONTIER

If we are going to establish permanent bases on the Moon or Mars, we need to know that they can be self-sufficient. Once we figure out what we're going to live in and how we're going to breathe, what are we going to eat?

Not to worry, because researchers in the Netherlands have managed to produce crops in Martian and lunar mock soils developed by NASA. They found that it is possible to grow food for future astronauts directly in the soil of another world and that the crops grown can produce viable seeds that can then be replanted.

The simulated soils were mixed with organic matter to provide nutrients for the crops, and standard Earth soil was used as a control sample. The team attempted to cultivate 10 different crops: garden cress, arugula, quinoa, tomato, radish, rye, spinach, chive, pea, and leek. The results show that it's only bad news for the Popeye fans among the astronauts, because spinach was the only crop that didn't reach a point where the scientists could harvest edible parts.

ACKNOWLEDGMENTS

4-5 (bkg) © Dima Zel/Shutterstock **4** (bl) © Gerald Robert Fischer/Shutterstock, (tr) © 3Dstock/Shutterstock, (br) © zenstock/Shutterstock **5** (c) gameover/Alamy Stock Photo, (tr) © Vadim Sadovski/Shutterstock, (cr) © Lotus Images/Shutterstock, (br) © Pedro Bernardo/Shutterstock **6-7** © Natalia Barsukova/Shutterstock, (bkg) © Alona_S/Shutterstock **8** © grebcha/Shutterstock **10-11** Bailey-Cooper Photography/Alamy Stock Photo **11** © rangizzz/Shutterstock **12** © Valentyn Volkov/Shutterstock **13** © Peter J. Traub/Shutterstock **14** © Aedka Studio/Shutterstock **14-15** © vilax/Shutterstock **16-17** Media Drum World/Alamy Stock Photo **18-19** © Jan Cejka/Shutterstock **20-21** © fizkes/Shutterstock **22-23** Public Domain {{PD-US-expired}} via Rutgers University Libraries and Wikimedia Commons **23** © Picsfive/Shutterstock **24** © Somchai Som/Shutterstock **24-25** © SakuraPh/Shutterstock **26** © Olha Birieva/Shutterstock **28** (bkg) © Luisa Fumi/Shutterstock, (c) History and Art Collection/Alamy Stock Photo **30-31** © Suchota/Shutterstock **32-33** © Hoiseung Jung/Shutterstock **34-35** © tomertu/Shutterstock **36** Public Domain {{PD-US-expired}} George III by Johan Zoffany via the Royal Collection Trust **38-39** © Denis Belitsky/Shutterstock **40-41** © 3Dstock/Shutterstock **42-43** (bkg) © Titima Ongkantong/Shutterstock **43** Ripley Entertainment **44** © Sean Locke Photography/Shutterstock **45** © Denise Kappa/Shutterstock **46-47** © Africa Studio/Shutterstock **48-49** © Olga Kot Photo/Shutterstock **50** © Sergey Nivens/Shutterstock **52-53** (bkg) © aleks.k/Shutterstock, (l) © Potapov Alexander/Shutterstock **53** (c) © Arina_B/Shutterstock **54** Harris & Ewing Collection, Prints & Photographs Division, Library of Congress, LC-H2- B-7464 **55** (bkg) © Lukasz Szwaj/Shutterstock, (c) gameover/Alamy Stock Photo **56-57** © Davide Calabresi/Shutterstock **56** (l) © KREML/ Shutterstock **58-59** (bkg) Dave Watts/Alamy Stock Photo **59** (tl) © Parilov/Shutterstock, (bl) © FamVeld/Shutterstock, (bc) © jgorzynik/Shutterstock **60-61** (bkg) © argus/Shutterstock, (b) © CHIARI VFX/Shutterstock **62** (b) © Gjermund/Shutterstock **62-63** © Pedro Bernardo/Shutterstock **64-65** {{PD-US-expired}} Accessed via Wikimedia Commons and Detroit Institute of Arts **66** © Tau5/Shutterstock **68** Public Domain {{PD-USGov-NASA}} **70-71** David Stephenson/Lexington Herald-Leader/Tribune News Service via Getty Images **72-73** © Everett Historical/Shutterstock **74** © Andrey_Popov/Shutterstock **77** Courtesy of U.S. Department of Energy **78-79** © HikoPhotography/Shutterstock **80** © xpixel/Shutterstock **81** © Gerald Robert Fischer/Shutterstock **82-83** © kudla/Shutterstock **83** © AG-PHOTOS/Shutterstock **84** © Vadim Sadovski/Shutterstock **86** (bkg) © Sergey Nivens/Shutterstock, (c) © robypangy/Shutterstock **88-89** Morgan collection of Civil War drawings; Waud, William, -1878; Prints & Photographs Division; Library of Congress; LC-DIG-ppmsca-21750 **91** © Marco Rubino/Shutterstock **92** {{PD-US-expired}} Accessed via Wikimedia Commons and Beinecke Rare Book & Manuscript Library, Yale University **92-93** © lunamarina/Shutterstock **95** © Oleksandr Kulichenko/Shutterstock **96-97** © Levent Konuk/Shutterstock **98-99** © Andrea Izzotti/Shutterstock **100-101** Mother Shipton's Cave **102** (bkg) © DmitriyRazinkov/ Shutterstock, (c) © likekightcm/Shutterstock, (b) © JLwarehouse/Shutterstock **104-105** © krungchingpixs/Shutterstock **106-107** © Romolo Tavani/ Shutterstock **107** © Astor57/Shutterstock **108-109** © antony cullup/Shutterstock **110-111** (bkg) © Deni_Sugandi/Shutterstock, © WhiteBarbie/ Shutterstock **112-113** © Herschel Hoffmeyer/Shutterstock **114-115** ADIFO/Cover Images **115** © Javier Rosano/Shutterstock **116** (bkg) © Gill Copeland/Shutterstock **116-117** Universal History Archive/UIG via Getty images **119** (bkg) © Dima Zel/Shutterstock, (c) © Leo Blanchette/ Shutterstock **120-121** Timothy Fadek/Corbis via Getty Images **122-123** CC BY 2.5 by frankenstoen **124-125** © SkillUp/Shutterstock **126** © Brian A Jackson/Shutterstock **128-129** © maxuser/Shutterstock **130-131** © Gorodenkoff/Shutterstock **132-133** © gan chaonan/Shutterstock **133** © Photoongraphy/Shutterstock **134-135** © zakalinka/Shutterstock **136** © Everett Historical/Shutterstock **136-137** © Lia Koltyrina/ Shutterstock **138-139** Public Domain {{PD-USGov-NASA}} **140-141** Design Pics Inc/Alamy Stock Photo **142** © doomu/Shutterstock **143** CTK/ Alamy Stock Photo **144-145** © Hung Chung Chih/Shutterstock **145** © SOMMAI/Shutterstock **146-147** (bkg) © Triff/Shutterstock **146** (bl) © koya979/Shutterstock **147** © Timofeev Vladimir/Shutterstock **148-149** © Pixelcruiser/Shutterstock **150-151** © Oleksandr Bilchuk/Shutterstock **153** © imging/Shutterstock **154-155** EFE News Agency/Alamy Stock Photo **156-157** University of Queensland/Cover Images **158-159** © D. Kucharski K. Kucharska/Shutterstock **159** © Mriya Wildlife/Shutterstock **160** (bkg) © Zakharchuk/Shutterstock, (c) © Fer Gregory/Shutterstock **161** © dmitro2009/Shutterstock **162-163** (bkg) © azazello photo studio/Shutterstock **163** (l) © Jagodka/Shutterstock **164-165** © Africa Studio/ Shutterstock **166** © Lotus Images/Shutterstock **166-167** IJF BURNS UNIT/CATERS NEWS **168** (bkg) © fotorath/Shutterstock, (c) © Viorel Sima/ Shutterstock **170-171** Public Domain {{PD-USGov-NASA}} Photographer: Robert Markowitz **172-173** Claudio Contreras via Minden Pictures **174-175** © Gorodenkoff/Shutterstock **176-177** © kikujungboy/Shutterstock **178** © HQuality/Shutterstock **180-181** © ugurv/Shutterstock **181** © stockphoto-graf/Shutterstock **182-183** © angela Meier/Shutterstock **184-185** (bkg) Public Domain {{PD-US}} Library of Congress Prints and Photographs Division Washington, D.C. 20540 USA http://www.loc.gov/pictures/item/00652085/, (b) © Alones/Shutterstock **185** (l) © Philip Arno Photography/Shutterstock **186** © seeyou/Shutterstock **186-187** © UfaBizPhoto/Shutterstock **188-189** © Willyam Bradberry/Shutterstock **189** © Elnur/Shutterstock **190-191** © Gorodenkoff/Shutterstock **191** © Zeeking/Shutterstock **Master Graphics** Courtesy of IFL Science

Key: t = top, b = bottom, c = center, l = left, r = right, bkg = background

Every attempt has been made to acknowledge correctly and contact copyright holders and we apologize in advance for any unintentional errors or omissions, which will be corrected in future editions.